Y0-AAG-918

Spare Time Guides, No. 6

SPARE TIME GUIDES SERIES

Ann J. Harwell, Editor

No. 1 *Automotive Repair and Maintenance*. By Robert G. Schipf.

No. 2 *Hunting and Fishing*. By Cecil F. Clotfelter.

No. 3 *Home Repair and Improvement.* By Robert G. Schipf.

No. 4 *Crafts for Today: Ceramics, Glasscrafting, Leatherworking, Candlemaking, and Other Popular Crafts.* By Rolly M. Harwell and Ann J. Harwell.

No. 5 *Stamps and Coins*. By Richard H. Rosichan.

No. 6 *Wine, Beer and Spirits*. By Dean Tudor.

SPARE TIME GUIDES:
Information Sources for Hobbies and Recreation, No. 6

Wine, Beer and Spirits

DEAN TUDOR

1975
Libraries Unlimited, Inc.
Littleton, Colorado

Printed in the United States of America

Library of Congress Card Number 74-80964
International Standard Book Number 0-87287-081-2

LIBRARIES UNLIMITED, INC.
P.O. Box 263
Littleton, Colorado 80120

CONTENTS

PREFACE

The Spare Time Guides series was conceived to provide librarians with selective, annotated lists of recommended books on specific hobbies and recreational activities, and also to help craftsmen and do-it-yourselfers learn more about their hobbies and crafts.

Because of the current interest in leisure time activities—and the resultant increase in the number of books concerning these activities—it has become difficult for librarians and hobbyists to determine just which books will fit their specific needs. Few selection aids on hobbies are available—the reviewing media that serve librarians devote little space to books on crafts and leisure time activities, and a search of hobby and crafts magazines reveals only relatively few book reviews. In too many cases, the only source of information on crafts books is the promotional material provided by the publishers, which does not usually offer a critical evaluation of the work. The Spare Time Guides series is designed to provide sufficient information to enable librarians and hobbyists to distinguish between books of varying quality.

Recent emphasis on appreciating and respecting wine's subtleties has made many people aware of the existence of a wine literature; this guide will serve a wide audience of amateurs seeking to expand their knowledge. Not only amateur oenologists, however, but anyone investigating the finer points of drinking will find ample critical guidance here. *Wine, Beer and Spirits* lists and annotates 494 books and 41 periodicals on all aspects of the subjects, with additional chapters that provide information on associations and clubs, and on museums, libraries, and contacts. The "directories" chapter gives addresses for all publishers whose works are listed, for sources of out-of-print materials, and for supply sources for wine- and beer-making equipment.

Subjects that are tangential to the main topics of wine, beer, and spirits are covered in detail: the reader will find books on cocktails, cooking with wine, health aspects of drinking, historical works, literary and musical references to drinks and drinking, and tasting guides, among other subjects. Do-it-yourself books for the home brewer or the home vintner are listed and described in a separate chapter. For Chapter 6, the author has carefully ferreted out and evaluated elusive audiovisual materials.

Dean Tudor is Program Director of the Library Arts Department at the Ryerson Polytechnical Institute, Toronto. Like many of us, he has long had an interest in fermentations and the pleasures that result therefrom; we can only be grateful that his interest has come to fruition in this thorough collection of materials on wine, beer, and spirits.

Ann J. Harwell
Editor, Spare Time Guides

INTRODUCTION

Throughout history, man has been attracted to alcoholic beverages for far more than just consumption. To many, winemaking is an economic necessity: a job. More than one-fifth of all the agricultural workers in the world depend on the grape. Beer (the easiest alcoholic beverage to make) is vitally needed in those areas where the water is usually not safe for regular consumption. And among the other uses of spirits, the results from the discovery of distillation have assuaged more hot tempers than they have created. That many troubles of the world can be attributed to alcoholic beverages cannot be denied. Nevertheless, the benefits easily outweigh the negative factors. Prestigious champagne is thought of as being symbolic of the good life to which we all aspire; the Christian faith equates (or very nearly equates, depending on the sect) red wine with blood. The first French Republic even named a month for the harvest season: Vendémiaire. And where would we be without our stockbroker, who derived from the French *broquier*—a man who tapped or broke a cask to draw wine (he was the only one allowed to do so, for he was the guarantor of quality demanded by both the merchants and the growers).

Obviously, this book will stress the positive benefits of alcoholic consumption. There is nothing here on "alcoholism" or social diseases. The appreciation of wine, beer, and spirit goes beyond knowledge of bottle contents and consumption. These are just the tip of the prodigious iceberg. An exhaustive evaluation of alcohol appreciation reveals that it can be a full-time study—alcohol appreciation contains all the components of a hobby.

The possibilities are limitless. The cultivation of a personal vineyard, such as Dr. Konstantin Frank's in up-state New York, can take up all of one's waking hours from spring to fall. Making wine, beer, or liqueurs at home can give immense creative satisfaction. Touring manufacturing plants, châteaux, vineyards, and quiet little towns in search of romantic lore makes a perfect holiday and good travelling both at home and abroad. Building a wine cellar or a wine closet in the home is a constructive task for those who like to work with their hands. The traditional "crafts and collecting" aspects of hobbies is exemplified in the collecting of old bottles or soaked-off wine labels, buying precious stemware and other drinking vessels (or making them from cast-off bottles), and acquiring beer mats, menus, matches, toasts, and serviettes. "Tastings"–either with a single wine-lover at home and at the vineyard, or with special friends at special parties–provide a festive element to the hobby. This leads naturally to cooking with alcoholic beverages, and to entertaining at home, and perhaps to forming a club or society for affiliation with a larger, internationally organized group such as the International Wine and Food Society.

For those of a scholarly bent, there is, of course, just the simple reading of books and articles, perhaps with a view to starting one's own personal library. The historical excursions are endless, and can be supplemented by many locally offered courses through a secondary school or community college. Night courses from extension departments are offered in the areas of home entertainment, wining and dining, cookery, wine appreciation, and amateur wine- or beer-making. For a listing, consult either the nearest public library or your state education office. Below is a list of addresses that may prove helpful for the avid "student" who wishes to pursue scholastic, rigorous studies in wines, beers, and spirits, with perhaps an ultimate goal of contributing something to the field. All of these groups offer serious studies that are available by correspondence course.

University of California at Davis
School of Viticulture and Enology
Davis, California

California Wine Advisory Board Study Course
Department W 3
717 Market Street
San Francisco, California 94103

The Beverage School (formerly the Grossman Course)
c/o Professor Henry Barbour
132 East 35th St.
New York, New York 10016

The Four Seasons Group
14 East 60th Street
New York, New York 10022

New York City Community College
c/o Tom Ahrens
300 Jay Street
Brooklyn, New York

The Connecticut Center for Continuing Education
East State Street
Westport, Connecticut 00680

Vintage Magazine Seminars in Wine
P.O. Box 851
New York, New York 10010

The Sommelier Society of America
121 West 45th St.
New York, New York 10036

And last, but not least, for those who already have a heavy interest in the fine arts (e.g., photography, painting, music), there is the possibility of making alcoholic beverages the subject matter—tavern drinking songs, a painting of food and wine, photographs of an Alsatian vineyard or Roquewiehr.

That wine, beer, and spirit manufacturers are enjoying their best years ever cannot be disputed. The amount of moonshine production in the United States continues to drop dramatically since more effective policing procedures were instituted five years ago. The public demand for lighter drinks means that a more efficient, higher proofing distillation process can be used, so that spirits can be quickly and cheaply produced in greater quantities. Recent advances in brewing have produced an acceptable draft-in-the-bottle substitute for those drinkers who prefer an unpasteurized taste. Ale is once again being brewed in the northeastern United States; if this experiment succeeds, ale will be nationally distributed.

In 1972, 332 million gallons of domestic wine were sold. By 1982, this total is expected to reach 662 million gallons, or three gallons per capita—still well below the European consumption rates. Gallo Brothers (125 million gallons) and United Vintners (50 million gallons) account for over half of the total market; indeed, fully 83 percent, or 239 million gallons, of the 1972 sales came from California. Production of California dry table wines was up 39 percent over 1971, while dessert wines were up only 13 percent—indicating the increasing popularity of drier wines.

The United States alone accounts for 60 percent of Bordeaux's choicest and most expensive wines. The 1972 imports from here totalled 16 million dollars. Italy, France, Portugal, Spain, and West Germany (in that order) provide us with 92 percent of our total wine imports. To handle these dramatic increases in both domestic production and wine importing, there have been several industrial mergers and corporation takeovers (such as Pepsi Cola's recent acquisition of wineries). Indeed, spirit manufacturers such as Heublin and Seagram's have been in the forefront of the wine industry takeovers.

Needless to say, many pressing problems have resulted from the increased demands for a limited supply of wine. And in many cases the supply itself is dwindling. On a pro-rated basis since 1900, there have been exceptionally bad or mediocre "wine years" between 1965 and 1973–not only in the "premiers crus" but also among the "vins ordinaires." In addition to climate, politics plays a negative role. The political situations in Israel, Greece, and Chile have adversely affected the harvest. Human incompetence is another sad area. A 1973 fire in Burgundy eliminated a half million cases of wine.

At the same time, interest in wine consumption has increased, bringing with it higher prices for the notable wines and raising the opening prices for the rest. And some countries, like Germany, have at the same time tightened their wine laws to reduce previously marketable high quality wines that presumably were not high quality in 1971/72. These laws are also difficult to administer.

A number of events have transpired, though, to ensure adequate supplies of wine. Domestically, new wineries are always opening in California, and indeed, one-fifth of the total 1973 acreage in that state did not exist prior to 1970 (although it should be stated that the choicest lands were furrowed long ago). Wineries are also introducing new types of wine, such as the immensely successful "pop" wines and hitherto neglected varietals. Wines from concentrates (usually Spanish) are also being made commercially. The acreage of A.O.C. Champagne lands was nearly doubled by a 1972 French regulation. A.O.C. Beaujolais is constantly expanding, while new markets are being explored for Argentinian, Yugoslavian, Bulgarian, and Hungarian wines. China shipped its first wines to the United States in 1973.

On the treacherous side, in 1969 certain Italians conspired to produce a "red" wine made entirely of ox blood, chemicals, and alcohol. When the evidence (over 10,000 gallons) was impounded it was shortly "stolen" by use of a siphon. In 1972 a similar situation occurred with Italian white wine that was hastily removed from the market. The French government in 1973 investigated rumours of watering down red Bourdeaux and Burgundies by the addition of white wine, thus casting aspersions on the reliability of even the "premier crus." At home, the federal government has banned the import of Spanish wines bearing French place names such as Chablis, Sauternes,

Burgundy—even though domestic manufactures are allowed to continue to use these appellations.

Wine is not alone with its problems—a worldwide shortage of barley, hops, and malt is affecting the beer industry; American distillers diverted much energy and capitol toward the production of "light white" whisky, then suffered the double blow of not having enough regular whisky on stock and abysmal sales of the lights; the shortage of heating oil over the 1973 British winter has forced cutbacks in the production of Scotch.

Wine literature is very well represented and well organized, more so than the literature for beers and spirits. The quality of writing is quite high, especially as the *literati* have taken to producing many lines about wine. Perhaps beer does not appeal to their literary palates. That there are so few materials on spirits is surprising, but perhaps this can be partially explained by the fact that there is little romanticism involved—there are no "vineyards" to visit, too many mechanical processes, and so few manufacturers (compared to the number for wine); plus the fact that home production of it is illegal. It is difficult to get excited about something that one cannot produce or touch or see. Materials are exceptionally scarce on liqueurs, infusions, and bitters (particularly the latter), probably because of the secrecy of the recipes used to blend and produce them. The fact that they are less used in cooking is also a detriment. Consequently, the present volume emphasizes those passages in books that cover liquor.

When looking for reading materials on wines, beers, and spirits (either to update this book or to search within the indexes and bibliographies mentioned in this book), the user should keep in mind library accepted forms of subject descriptors and indicators. Various indexes and the Library of Congress subject headings were consulted to produce the lists which follow. All of the descriptors are stated in upper case characters. These descriptors, which are in common usage, will assure effective retrieval of the maximum number of entries containing the sought-for information.

WINE. The process is FERMENTATION, and general material usually appears under the heading WINE AND WINEMAKING, with a geographic breakdown and a special classification for AMATEUR. More specific material can be found under the name of the country, or APERITIFS, BORDEAUX, BURGUNDY, CHAMPAGNE, CHIANTI WINE, CLARET, FRUIT WINES, HOCK (WINE), MADEIRA WINE, MOSELLE WINE, PORT WINE, SAUTERNES, SHERRY and SPARKLING WINES. Additional material can be located by checking entries under COOKERY (WINE), GRAPES, VITICULTURE, and the heading WINE IN (art, architecture, literature, music, etc.).

BEER. The process is FERMENTATION and BREWING, and general material usually appears under the heading BEER. More specific material can be found under the general headings of ALE, BREWERIES, CIDER, MALT

LIQUOR, LAGER, PERRY. Additional references can be located by checking entries under COOKERY (BEER) and MALT.

SPIRITS. The process is DISTILLATION, and general material usually appears under the heading LIQUORS. More specific material can be found under the type of liquor headings of BRANDY, COCKTAILS, GIN, LIQUEURS, RUM, VODKA, WHISKY (Bourbon, Canadian, Scotch). Additional references can be located by checking the entries under DISTILLING INDUSTRY, DRINKING IN LITERATURE, and HOTELS, TAVERNS, ETC.

As to the format or organization of this multi-media bibliographic (or is it mediagraphic?) guide, the table of contents is self-explanatory. There is a full author-title-subject index. Availability and prices are given in all cases where known; there is a directory of publishers given at the end of this book. Chimo!

January 1974

Dean Tudor
Toronto, Canada

Chapter 1

REFERENCE WORKS, TECHNICAL LITERATURE, AND GENERAL WORKS

REFERENCE WORKS

GENERAL

This section includes notable consultative sources, statistics, atlases, dictionaries and encyclopedias. Many of these works are indispensable. Subsections beginning on page 24 and page 26 provide information on directories and bibliographies, respectively, for those who wish to pursue the subject.

1. **Alcohol and Tobacco Summary Statistics**. 1953– . Washington, D.C., Government Printing Office. annual. $1.00.

Statistics cover U.S. production, withdrawals and stocks of distilled spirits, wine, beer and tobacco, with comparative data by states and by months. There are historical tables and over 90 statistical tables.

2. **Alcoholic Beverage Industry Annual Facts Book**. 1946– . New York, Licensed Beverage Industries, Inc. annual. free.

As a promotional work, the text concerns the role of the alcoholic beverage industry in the national economy and in the social and cultural life of America.

3. **Annual Survey of Illegal Distilling in the United States**. 1946– . New York, Licensed Beverage Industries, Inc. annual. free.

This is a compilation of official statistical data on illegal moonshining operations in America. It deals with the medical, social, and economic effect of these operations on the nation.

4. **Beverage Media Blue Book**. 1935– . Beverage Media, Ltd., 251 Park Avenue South, New York, 10010. annual. $6.00. LC 51-38397.

This business publication focuses on product facts, product knowledge, trade education, and rules regulating alcoholic beverages. The directory section covers New York only.

5. **Brewers Almanac.** United States Brewers Association, Inc., 1750 K. Street, N.W., Washington, D.C. 20006. annual. $15.00. LC 45-51432.

Contains statistics of production, withdrawal, taxes, exports, labor, consumption, retail outlets, Repeal, and the local options in the states, mainly from the latest available U.S. Census of Manufactures data. Information is historical, with the retrospectives given by region or state.

6. **Brewers Almanack**. 1888– . 19 Briset Street, London, EC 1, England. annual. $15.00.

With much of the same material as in its American counterpart, this publication covers the United Kingdom and Eire.

7. Comité National des Vins de France. **Premier annuaire des caveaux, celliers, chais, et autres centres de dégustation des vins de France**. Paris, 1971. free.

This first annual, comprising 127 pages, is a listing of all groups involved with wines in France (growers, chateaux, co-operatives, merchants, shippers, cellars, special wine tasting sports), divided by wine producing area and subdivided by Department and then by town within. Names, addresses, telephone numbers, opening hours, and some indication of tastings are set out. This is a major book that the Comité spent some time on, and it is an essential possession for the traveller in France. In addition to wines, brandies (such as cognacs and armagnac) are also covered.

8. Debuigne, Gérard. **Larousse des vins**. Paris, Larousse, 1970. 272 p. illus. (part col.). 86 Fr.F. LC 73-867908.

At present available only in the French language, this comprehensive compendium of wine lore and hard core information could prove to be a major and standard reference source, measuring up to other Larousse texts. Dictionary arrangement.

9. Distilled Spirits Institute, Inc. (Washington, D.C.). **Annual Statistical Review**, 1951– . annual. free. LC 54-17193.

Provides analysis of the beverage distilling industry, with highlights of the local option elections each year. There is a tabulation of the "wet" and "dry" population of all states as of December 31 of each year. Comparative and historical data go back only to Repeal. Other tables include public revenues, tax rates, import duties, stock, bottled output and foreign trade.

10. Distilled Spirits Institute, Inc. (Washington, D.C.). **Summary of State Laws and Regulations Relating to Distilled Spirits**. 1935– . biennial. $7.50. LC 36-8319 rev.

The twentieth edition (1972) covered in its 92 pages a great deal of material, all of it compiled from state statutes, administrative regulations, interpretative rulings, and replies to questionnaires. Part One concerns the control (monopoly) states, with tabular data and explanatory notes. Part Two is about the licensed states, and contains similar data. Miscellanous tables are in Part Three, and comprise fees (by state), commodities other than distilled spirits which may be sold (e.g., potato chips, tobacco) and a complete list of federal excise tax rates on distilled spirits since 1791, when it was 9 cents per gallon (now, it is $10.50).

11. Distilled Spirits Institute, Inc. (Washington, D.C.). **Public Revenues from Alcoholic Beverages**. 1937– . annual. free. LC 38-17702.

A straightforward accounting, with the appropriate tables, of the tax money derived from the sales of alcoholic beverages, operation of state liquor stores, and license fees, with comparative data and type of beverage sold. Compilation is by both federal and state data. Reports show the method of controlling local collections, and allocation of state taxes, where available.

12. Distilled Spirits Institute, Inc. (Washington, D.C.). **Retail Outlets for the Sale of Distilled Spirits**. 1971. 101p. charts, free. LC 54-43588.

A compilation, by state, with brief population summaries to indicate the per capita service given by the stores (on- and off-premises).

13. Doxat, John. **The World of Drinks and Drinking: An International Distillation**. New York, Drake, 1972. 256p. illus. bibliog. $8.95 LC 72-1340.

This is an encyclopedia with entries in alphabetical order, with appropriate cross references where needed. Through it all, Doxat examines the drinking patterns of major nations around the world (e.g., the minimum drinking age in Russia). Included are individual entries for bottles, closures, toasts, customs, distilling, and brewing–to name but a few. Anecdotes and quotations make this easy reading, especially the entry for "Vulgarity." Brief histories for all drinks are given. What makes this book especially valuable are the short corporate histories for wine merchants and distilleries. There is also a notable de-emphasis on wines and a subsequent upgrading of liquor.

14. **Elsevier's Dictionary of Barley, Malting, and Brewing, in Six Languages: German, English/American, French, Danish, Italian, Spanish.** Comp. and arranged on a German alphabetical base by B. D. Hartong. With a preface by Ph. Kreiss. Amsterdam and New York, Elsevier Publishing Co.; distr. Princeton, N.J., Van Nostrand, 1961. 669p. $25.00. LC 60-12356.

As one of Elsevier's authoritative multilingual dictionaries, this is a very helpful translation dictionary.

15. **European Spirits.** 1970– . Stamex, B.P. 505, Hilversum, Pays-Bas, Belgium, triannual. 900 Belgian Fr.

The second edition, for 1973 to 1975, covered in its 264 pages the European distillers of whiskey, gin, eau-de-vie, vodka, cognac, schnapps and liqueurs. The directory of 1,100 businesses in 25 countries details: name and address; telephone, telex and/or telegraph numbers; year founded; number of employees; names of directors and export heads; affiliated groups; and production programs. There is also a register of 1,650 trademarked products.

16. **From the State Capitals: Liquor Control.** 1946– . Bethune Jones, 321 Sunset Avenue, Asbury Park, N.J. 07712. weekly. $72.00 per year.

This processed periodical covers state and local regulations throughout the United States affecting the production and marketing of alcoholic beverages: one of a series of reporting services.

17. Grossman, Harold J. **Grossman's Guide to Wines, Spirits and Beers.** 4th rev. ed. New York, Scribner's, 1964. 508p. illus. maps, bibliog. $8.95. LC 64-24895.

A standard reference book, not arranged alphabetically but by topic. Short introductions deal with definitions and fermentation, then the body swings into the wine-producing countries of Europe. There are 50 pages on America, material on distilled spirits, but only 14 pages on beer. Recipes are also included for cocktails and for food preparations utilizing wine. Although this book has sold well to the general public, it was written for the hotel and food industry. This fact accounts for its plodding style and the inclusion of material dealing with the wine list, menu making, rules for bartending, hotel and restaurant glassware, inventory, merchandising and accounting procedures, plus a section on the stemware that Grossman himself designed as all-purpose, with an abominable etched line at the 1 oz., 2 oz., or 4 oz. tide lines: Haszonics (see entry 112) is better for this information. Numerous appendices include a quick guide to wines and spirits, cost and profit charts, wine classification, a directory of American wine and spirit producers and importers, a glossary, and an extensive index. The book needs to be revised, since it is now a decade old.

18. James, Walter. **Wine: A Brief Encyclopedia.** New York, Knopf, 1960. 208p. illus. $3.75. LC 60-50601.

Published in England a year earlier as *A Word-Book of Wine*, this short account takes in the worldwide vineyards of repute. There are names of wines, grape varieties, vineyards, wine-growing areas, bottle sizes, alcohol strength, the proof system, and matters dealing with wine snobbery.

19. Johnson, Hugh. **The World Atlas of Wine: A Complete Guide to the Wines & Spirits of the World**. London, Mitchell Beazley; distr. New York, Simon and Schuster, 1971. 272p. illus. (part col.). index. $25.00. LC 71-163481.

Johnson's world atlas is something more than an atlas, since it includes basic information on winemaking and choosing wines. At the same time, it is less than what the subtitle promises: a complete guide to the wines and spirits of the world. The book, printed on heavy paper and lavishly illustrated with color and black and white photos, charts, reproductions of numerous wine labels, and 143 well-drawn, detailed maps, is divided into seven parts: introduction (history of wine, winemaking, etc.); choosing and serving wine (with vintage charts); France; Germany; Southern and Eastern Europe and the Mediterranean; the New World; Spirits. An index and a 7,000-entry gazetteer are appended. France is given as much space (74 pages) as that assigned to Germany and the rest of Europe, while all other wine-producing areas (United States, Australia, South Africa, South America, England, and Wales) are compressed into 26 pages. With this allotment, even France is not covered completely, let alone any other part of the wine world. The 19-page section on spirits is too brief to be of much value and seems out of place. A double page is allotted to each region, giving a survey of the region, the grapes grown and major wines produced, and other descriptive matter; drawings of selected labels; a locator map, photo of the area, and a detailed map. These maps are the focal point of the work. Instead of showing only political boundaries, they have viticultural detail, indicating contours, elevations, vineyards, and woods. Included in the two introductory sections are some interesting maps of the world's vineyards by countries, in thousands of hectares; graphs depicting world wine consumption, by country; and data on wine producing areas in the ancient world and the Middle Ages. Among the several informative pieces of miscellany are a double-page drawing of the layout of a chateau in all its detail and a brief selected bibliography on wine. This "atlas" differs from other wine books in its emphasis on viticultural and economic detail rather than on the simple geographical location of chateaux. It is a useful, albeit expensive, supplement to Lichine's *Encyclopedia of Wines and Spirits* (Knopf, 1967).

20. Larmat, Louis. **Atlas de la France Vinicole**. Paris, L. Larmat, 1941– .

The only other wine map book is Johnson's *World Atlas of Wine* (see entry 19). The Larmat atlas provides the definitive series of maps for French vineyards–invaluably detailed. Prepared under L'Institut National des Appellations d'Origine and published in parts, this is a work in French that covers: *Les vins de Bordeaux* (1949); *Les vins de Bourgogne* (1955); *Champagne* (1944); *Les vins des Côtes du Rhone* (1943); *Les vins des Côteaux de la Loire*; and *Les eaux-de-vie de France: le cognac*.

21. Layton, T. A. **Winecraft: The Encyclopedia of Wines and Spirits**. Rev. ed. London, Harpers, 1959. 295p. £2; pa. £1.

This work was originally serialized in *Harper's Wine and Spirit Gazette* (1957-1959) and is a complete revision of the 1935 edition, which was the first book of its kind then to be published in English. It is more a dictionary than an encyclopedia, with short entries in alphabetical order. The appendices contain: vintage charts of Burgundy, Champagne, Bordeaux, Cognac, Moselle, Rhine, Port, Sherry; glossary of terms; and an invaluable list of 23 common phrases in English, French, Spanish and German, such as: "How has the rain affected the grapes?" and "We do not ship any wine in cask, only in bottle."

22. **Lexique de la vigne et du vin**. Paris, Office International de la Vigne et du Vin, 1972. 700p. maps. 100 Fr. F.

This is a translation/terminology dictionary for wines like Elsevier's similar work for beer. The seven languages are English, French, Italian, Spanish, German, Portuguese, and Russian. A great number of specialists worked on this project for many years, and the resulting dictionary is in four parts: 1) 2,000 terms and definitions (510 pages); 2) seven alphabetical indexes (one per language) plus a special section for Latin terms cited in the book; 3) exhaustive units of measures (both past and present) and tables of equivalencies; and 4) 17 maps and drawings. This is an essential work for the serious student of wines.

23. Lichine, Alexis. **Encyclopedia of Wines and Spirits.** In collaboration with William Fifield and with the assistance of Jonathan Barlett and Jane Stockwood. New York, Knopf, 1967. 713p. maps. LC 66-19385. $17.50.

Wine producer and wine merchant Lichine has here produced *the* definitive basic information book about wines, beers, and spirits. His superb and well-written introductory material covers history; wine, food, and health; wine cellars; vinification processes; viticulture; and spirit making. The main body itself is alphabetical in arrangement and self-indexing. Lichine covers geographic areas with the types of vines, types of wines, spirits, beers, aperitifs, and locally applied technical terms. Maps are of average display. Most entries are long, especially for French and German winegrowing areas, but at least the reader does not have to hunt around for chateaux or areas (such as Côtes du Rhone); these, while mentioned under the country (in this case, France) are given their own alphabetical entries. The "great area" entries have lists of *crus;* an important feature here is the large number of Bordeau chateaux and Burgundy vineyards described in the text under their own entry. Appendices include: classifications of Médoc wines, plus Pomerol, Graves, and St. Emilion; container information; tables of spirit strength; conversion tables; vintage charts; pronouncing glossary; and a good historical bibliography. There is little on beer, but the spirits seem to be well represented. A well-organized book that is very enjoyable to read (unlike most reference sources), this book needs revision only in those charts where dates serve an important function. A revised edition is planned for late 1974.

24. **Liquor Handbook**. 1954– . Gavin-Jobson Assoc., Inc., 488 Madison Avenue, New York, 10022. annual. $14.50; $12.00pa.

This handy compendium provides market trend information and statistics that provide the basis for marketing and advertising decisions of distillers, vintners, and importers of alcoholic beverages. Each annual has around 350 pages of tables, maps and projections, and ads. The four sections cover: 1) The National Liquor Market (consumption, sales, regional breakdowns, distribution maps, projections, retail licenses, taxation, bootlegging, and retail sales prices); 2) Distilling Operations (production, storage and aging, usage, and bottling); 3) The Market for Major Distilled Spirits Types (by type, including prepared cocktails, with maps and 15-year projections); 4) Advertising and Promotion (expenditures, directory of media personnel, the black market, outdoor advertising, and packaging).

25. Marcus, Irving H. **Dictionary of Wine Terms**. 16th ed. Berkeley, Calif., Wine Publications, 1972. 72p. $0.75pa.

As a long-time editor of *Wines and Vines*, Marcus has had first-hand experience with the industry. Here he presents the kernel of his experience in a short, easily portable book dealing with the subject of wines but *not* any types or vines. He defines, carefully and clearly, 600 terms used in the wine industry. He also outlines the winemaking process and includes data on wine production around the world. Primarily because of its use in the trade and in the classroom, 350,000 copies have been sold worldwide.

26. **Memento de l'O.I.V.** 1970 ed. Paris, L'Office International de la Vigne et du Vin, 1970. 1211p. 65 Fr. F.

This guide, in French, cuts a clear path through the voluminous documentation about governmental sources and legislation. There are five main chapters:

legislation (890p.) from 40 countries (wine codes, texts of laws, new laws, etc.) plus legislation of the E.E.C. generally; statistics (190p.) on cultivation, production, importation, exportation, consumption of all grape products (wine, table grapes, raisins, grape juice) for 70 countries, along with historical figures; *appellations d'origine* (55p.) which lists wines and principal growths for 20 countries, along with the area determination laws; a wine periodicals list that furnishes the names and addresses of 140 periodicals in 27 countries; and a list of national associations of wine and grape-growing in 40 countries. A *very* useful book.

27. **National Association of Alcoholic Beverage Importers**. Annual Statistical Report. 1961– . annual. free.

A series of tables to indicate what was imported into America, what was consumed, and where.

28. **Puerto Rico Rum Producers Association**. Statistics. 1943– . monthly, annual. restricted to members.

Production, bottling for United States, Puerto Rico, and foreign markets (cases and proof gallons), plus other alcoholic beverages of Puerto Rican manufacture and imports of alcoholic beverages into Puerto Rico.

29. **Répertoire des stations de viticulture et d'oenologie**. Paris, L'Office International de la Vigne et du Vin, 1972. 120p. 155 Fr. F.

This is useful information on 280 active research stations and laboratories dealing with viticulture or winemaking in 35 countries. Arrangement is by country, then alphabetically by name, with complete address, names of director and assistants, general program to be carried out, the manner of consultation, and working languages. An alphabetical index is included, plus a subject index to type (e.g., viticulture, winemaking, microbiology, table grapes, raisins, etc.).

30. Schoonmaker, Frank. **Encyclopedia of Wine**. 5th ed. New York, Hastings, 1973. 442p. illus. maps. $10.75. LC 77-9315.

This healthy competitor to Grossman is worldwide in coverage: unlike many wine "encyclopedias," it is always up to date, being constantly revised. It is more a dictionary, with expanded detail on France and California. There are over 2,000 wine names described, and over 100 maps and wine labels.

31. Simon, André L. **A Dictionary of Wines, Spirits and Liqueurs**. 1st American ed. New York, Citadel Press, 1963 (c.1958), 190p. illus. $4.95. LC 63-21203.

This is handy for snap definitions, but it is otherwise too general. For the traveller who wants nothing bulky but who needs more information than Marcus provides. (See entry 25.)

32. Simon, André L., and Robin Howe. **Dictionary of Gastronomy**. Rev. and enl. ed. New York, McGraw-Hill, 1970. 400p. illus. (part col.). bibliog. $15.95. LC 72-89318.

The first edition was published in 1949. This one has 2,000 definitions, 600 line drawings, and 64 full color illustrations. Most definitions include some historical references. There are complete discussions of wines and cheeses, as well as fruits, vegetables meats, fish, and fowl.

33. Simon, André L. **The International Wine Food Society's Encyclopedia of Wines**. New York, Quadrangle; distr. by Harper, 1973. 312p. maps. $15.00. LC 72-85052.

This is the legacy left behind after Simon died in 1970. It has been developed from a number of his previous books, such as *A Dictionary of Wines, Spirits, and Liqueurs* (see entry 31), itself based on a 1935 *Dictionary of Wine*, and his 1970 edition of *A Wine Primer*, often revised. His book opens with a brief 50-page description of wines of the world, followed by a 200-page "gazetteer" with 40 pages of maps. The 7,000 entries are either vineyard names or place names that correspond to types of wine; hence, great wines and *vins ordinaires* are given equal prominence. Entries are in alphabetical order, with adequate cross references. Each entry on the double-column pages describes (in about three lines): 1) the corresponding map number; 2) its correct designation in terms of its country's laws and regulations (e.g., France's A.O.C. "Pomerol"); 3) its location in terms of its largest neighbor; 4) a half-dozen words or so about its quality (e.g., "ordinary to fair light white dessert wine"); and 5) the country of origin. Its main value, then, lies in its quick distinction of one wine from others of a similar name. The quantity and brevity of the entries makes this book more a dictionary than an encyclopedia. It has no illustrations, and its maps are too general and too broad in coverage. Also, it would be useful to have a "reverse" dictionary available: wine descriptors listed, subdivided by the name of the wine available.

34. Time, Inc. **Language of Liquor**. New York, 1971. 28p. free.

A short but useful glossary of liquor terms and definitions, reprinted from *California Gold Book, Beverage Industry News.*

35. United States. Internal Revenue Service. **Regulations Under the Federal Alcohol Administration Act** (Title 27, Code of Federal Regulations). Washington, D.C., Government Printing Office. 1961. 70p. [IRS Pub. No. 449 (2-61)]. $0.30. LC 61-61159.

This is it, and it covers the non-industrial use of distilled spirits and wine, plus bulk sales, and the labeling and advertising of wines, beers, and spirits. Appended, by being tipped in, are Treasury Decisions 6597, 6601, 6672, 6776, 6799 and 6901, which affect the regulations slightly.

36. **Wine Marketing Handbook**. 1971– . Gavin-Jobson Assoc., Inc. 488 Madison Avenue, New York, 10022. annual. $8.50; $6.00pa. LC 72-626221.

This short publication (about 160 pages) does for wine what the *Liquor Handbook* does for spirits (see entry 24). Its set-up is very similar, being divided into four sections: 1) National Wine Market; 2) Production and Inventories; 3) Market for Major Wine Types (table, dessert, vermouth, champagne, and sparkling); and 4) Wine and Advertising.

DIRECTORIES

These works provide key information on personnel, addresses, statistics, etc.

37. **American Brewer Annual World Directory of Breweries**. $13.00. American Brewer Publishing Corporation, 33 Lyons Place, Mount Vernon, N.Y. 10553.

In addition to corporate names, there is other relevant information on key personnel and on production.

38. **American Brewer Register**. annual, $1.50 American Brewer Publishing Corporation, 33 Lyons Place, Mount Vernon, N.Y. 10553.

A listing of suppliers to the brewing industry by company names, addresses, key personnel, and product categories.

39. Anderson, Sonja, and Will Anderson. **Anderson's Turn of the Century Brewery Directory: A Complete Listing of All U.S. Breweries in Operation in 1890, 1898, 1904 and 1910, as Listed in the Original Brewers' Hand-Books for Those Years**. Comp. and published by Sonja and Will Anderson. Carmel, N.Y., 1968. LC 68-6732. lv. (unpaged). illus. price not reported.

40. **Annuaire des boissons.** 80 Fr. Fr. Editions du Gonfalon, Route de Dourdan, 91 Angerville, France.

An international directory to all drinks and the supplies industry. Published since 1945.

41. **Blue Book of Wine and Spirit Wholesalers**. annual. $10.00. Wine and Spirit Wholesalers of America, Security Building, St. Louis, Mo.

A directory to wholesale distributors of wines and liquors, with the names of company executives, listed by states. Also lists major suppliers, alcohol tax officials and associations.

42. **Bonded Wineries and Bonded Wine Cellars Authorized to Operate**. annual. free. Internal Revenue Service, Alcohol and Tobacco Tax Division, Washington, D.C. 20224.

43. **Breweries Authorized to Operate**. annual. free. Internal Revenue Service, Alcohol and Tobacco Tax Division, Washington, D.C. 20224.

44. **Brewery Manual and Who's Who in British Brewing**. annual. £2.50. Northwood Publications, Ltd., Northwood House, 93-99 Goswell Road, London, ECIV 7QA.

This is a comprehensive directory of the British brewing industry, listing all brewing companies, personnel, and financial information. The "who's who" section gives detailed background information on all individuals of importance in the industry: directors, head brewers, and bottling managers.

45. **Harper's Directory and Manual of the Wine and Spirit Trade**. annual. £2.50. Harpers Trade Journal, Ltd. 22 Cousin Lane, London, EC 4, England.

Section one, directory of wine and spirit trade in the United Kingdom and Eire (alphabetically); section two, a geographic arrangement of section one; section three, directory of wine and spirit trade overseas (alphabetically by country); section four, list of wines and spirits, shippers and agents, brand names, and ancillary trades (worldwide, arranged by beverage); section five, a list of trade associations and bonded warehouses in the United Kingdom and Eire; and section six, index.

46. **Modern Brewery Age Blue Book**. annual. $6.00. Modern Brewery Age Publishing Company, 80 Lincoln Avenue, Stanford, Conn. 06904.

Lists all breweries in the Western hemisphere, along with their annual production figures and the names of management and supervisory personnel.

47. **Red Book Encyclopedia Directory of the Wine and Liquor Industries**. Irregular. $22.50. Schwartz Publications, Inc. 6 West 57th Street, New York, 10019.

Much is covered here–publicly owned liquor companies; state and federal laws and regulations governing the industry; tax laws; U.S. wine and liquor producers and importers; state administration officials; a section on licensees; listings of wineries, wholesalers, equipment, packaging, and production methods; and an alphabetical index to alcoholic beverage brands.

48. **Shaws Wine Guide**. Three times a year. £2.00. Shaws Price Guide, Ltd., 4 The Broadway, London. N8 9SP.

This is one of Shaws' many industrial guides that are compiled with the cooperation of the British trade. Listed here are wines, beers, spirits, and ciders by their names, with the current list or retail prices. This gives an idea of the range available, and of the "recommended" prices. The strictly factual presentation never mentions selection or quality.

49. **Wine and Spirit Trade International Yearbook**. annual. $15.00. Haymarket Publishing Company, Gillon House, 6 Winsley Street, London, W1A 2 HG.

Formerly *The Wine and Spirit Trade Diary*, this book has four main sections: agents, shippers, distilleries, opening prices of Scotch whisky, distilleries (alphabetically by product), blenders and vintage summaries; merchants in the United Kingdom and Eire; whisky (blends and brands); and allied trade (customs, legislation, trade association lists).

50. **Wineries and Wine Industry Suppliers of North America**. annual. $5.00. Marcus Publications, 16 Beale Street, San Francisco, Calif. 94105.

Formerly *Annual Directory of the Wine Industry*, this item is the December issue of *Wines and Vines*. It contains data on all U.S. wineries and wine bottlers: size, officers, products, brands, distribution, buyer's guide. It presents a resume of laws and regulations affecting wine in each of the 50 states. Some coverage is also given to Canada and Mexico.

51. **The Wine and Spirit Trade Review Trade Directory**. annual. £2.50. William Reed, Ltd. 19 Eastcheap, London, EC 3, England.

This work with the double-barrelled title was formerly the *Directory of Companies and Their Subsidiaries in the Wine and Spirit and Brewery Trades*. Provides useful information on personnel and production.

BIBLIOGRAPHIES

Unfortunately, there appear to be few bibliographical works concerned with distilled spirits or beer. Readers interested only in either of these two types of beverage should go straight to Noling (entry 59).

52. Amerine, M. A., and L. B. Wheeler. **A Checklist of Books and Pamphlets on Grapes and Wine and Related Subjects, 1938-1948**. Berkeley, University of California Press, 1951. 240p. $5.00. LC 51-62205.

There are 1,789 items, with a bibliography of other bibliographies and full indexes. Annotations are short or partial.

53. Amerine, M. A. **A Short Checklist of Books and Pamphlets in English on Grapes, Wine and Related Subjects, 1949-1959**. [Davis], Calif. 1959. 61p. price not reported. LC 59-52146.

Produced with the assistance of the Davis Research Librarians for the 1959 annual meeting of the California Library Association.

54. Amerine, M. A. **A Checklist of Books and Pamphlets in English on Grapes, Wines and Related Subjects, 1960-1966**. Davis, Calif., 1969. 84p. price not reported. LC 77-14845.

Also includes a supplement for his 1959 publication, which covered 1949 to 1959.

55. Candwell, A. **Bibliographie des viroses de la vigne, des origines à 1965**. Paris, Office International de la Vigne et du Vin, n.d. price not reported. 76p.

Details material written on wine and wine diseases caused by viruses.

56. Dicey, Patricia. **Wine in South Africa: A Bibliography**. Cape Town, University of Cape Town Libraries, 1970. (University of Cape Town, School of Librarianship, Bibliographical Services). 23p. $0.50. Rand. LC 77-876454.

57. Dumbacher, Egon. **Internationale Weinbibliographie, 1955-1965 und Sachverzeichnis, 1956-1965 der Mittelunger Rebe und Wein**. Wien, Austria, Bundesministerium für Land– und Forstwirtschaft, 1966. Various pagings, 346 Sch. LC 68-80060.

Now largely continued by *Wein-Bibliographie* (see entry 62).

58. Lucia, Salvatore. **Wine and the Digestive System: The Effects of Wine and Its Constituents on the Organs and Functions of the Gastrointestinal Tracts, a Selected Annotated Bibliography**. San Francisco, Fortune House, 1970. 157p. $6.50. LC 74-123753.

The 500 entries are divided by subject (e.g., intestines, bladders, etc.) and arranged chronologically within these subjects.

59. Noling, A.W., comp. **Beverage Literature: A Bibliography**. Metuchen, N.J., Scarecrow Press, 1971. 865p. $20.00. LC 70-142238.

Covers 5,000 titles in total, about 1,200 of which are related to wine, winemaking, and grapes (mostly in English). Another 700 titles are concerned with beer, cocktails, cider and perry. Most of the balance deals with non-alcoholic beverages. The work includes: author list, subject list, description of subject categories, short-title list, appendices listing the reference sources consulted in this compilation, and a list of major libraries specializing in the field. Some annotations.

60. Simon, André L. **Bibliotheca Bacchia.** London, 1935. Reprinted Hopewell, N.J., Booknoll Farm, 1972. 2 v. in one. $50.00. Also reprinted by Alfred Saifer as Bibliotheca Bacchical Wine and Food Bibliography.

A listing of works published before 1600, concerning viticulture, the art of winemaking, table manners, drunkenness, decrees and regulations relating to the wine trade, rules of health, medical books and treatises on agriculture. Vol. 1 comprises incunabula, with 60 facsimile title pages among its 237 pages. Vol. 2 contains a description of 240 sixteenth century books that are not mentioned in Vicaire's *Bibliographie gastronomique*; and thus *Bacchia* complements it.

61. Simon, André L. **Bibliotheca Vinaria.** 1913. Reprinted by Finch Press, 1970. $24.00.

A bibliography of books and pamphlets dealing with viticulture, winemaking, distillation, the management, sales, taxation, use, and abuse of wines and spirits.

62. **Wein-Bibliographie: Deutschsprachiges Schriftum in Auswahl and Nachträge aus früheven Jahren.** Traben-Trarbach, Weinburg and Keller-Verlags, 1956– . LC 70-649056.

An annual publication covering the years from 1955 on. It is about two years behind in indexing. Introductions and captions are in English, French, and German.

TECHNICAL LITERATURE

For a proper understanding of wines, beers and spirits it is necessary to know something of the various components that go into the production of alcoholic beverages. Below is a selective list of technical texts that detail information about viticulture, harvesting, production of alcohol (fermentation and distillation), diseases and spoilage, plus general notes on composition and nutrients.

63. Amerine, Maynard A., and H. W. Berg. **Technology of Wine Making**. 3rd ed. Westport, Conn., Avi Pub-Co. 1972. 802p. illus. bibliog. $27.00 text ed. LC 78-188033.

First published in 1969, this is now a standard treatise on commercial wine making. The authors cover methods (composition, quality) used in all of the important wine regions of the world, with a special section on sherry processes: Bodega (Spanish), baking (California), and the rapid Tressler (New York and Ontario). New information here includes sections on wine yeasts (using the 1970 Lodder classification), winery design, equipment and operation; and the processing of vermouth and other flavored wines using herbs and spices, with a subsection on the California "pop" wines. Three chapters are concerned with spoilage, prevention, and waste disposal.

64. Amerine, Maynard A., and M. A. Joslyn. **Table Wines: The Technology of Their Production**. 2nd ed. Berkeley, University of California Press, 1970. 997p. illus. bibliog. $25.00. LC 69-12471.

This standard reference work is now two and a half times the size of its original edition in 1951. The massive changes in the past two decades through scientific discoveries in microbiology and biochemistry alone have shifted winemaking from an art to a science. World wine areas and wine types are described in detail, along with technical standards, automation, and instrumentation. Again we have a step-by-step guide to all aspects of winemaking–winery construction, marketing, sanitation, grape juice, fermentation processes, preservations, aging, filtration, and finishing. Other chapters cover the technical processes for red, rosé, white, sweet and sparking wines. Special sections deal with commercial wine disorders (for personal solutions, see the *Larousse gastronomique,* entry 387), summaries of research in wine tasting and evaluation, a 15-page selective bibliography and 117 pages of literature citations.

65. Amerine, Maynard A., and Vernon L. Singleton. **Wine: An Introduction for Americans.** Berkeley, University of California Press, 1965. 357p. illus. bibliog. $6.95; $2.85pa. LC 65-11785.

A factual account dealing mostly with the United States, since 90 percent of all types of wine consumed here is domestic. This is a popularly written lay guide to vinification practices (growing, fermentation, classifying, distilling). Other wines of the world are briefly mentioned. A first-rate book for the price, with an extensive 16-page bibliography.

66. Amerine, Maynard A., and A. J. Winkler. **California Wine Grapes**. Berkeley, University of California Press, Division of Agricultural Sciences, Agricultural Experiment Station, 1963. 83p. tables. (Bulletin 794). price not reported. LC 63-64007.

A short investigation into the composition and quality of California wine grape must and the subsequent wines produced. The prolific Dr. Amerine, who has been writing for 40 years, is America's top oenological scholar.

67. Austin, Cedric. **The Science of Wine**. New York, American Elsevier, 1968. 216p. illus. $6.75. LC 68-26815.

This is a basic science course for winemakers who want to know the reasons for changes taking place in the vat (for commercially made wines). There are four main sections: yeast, sugar, alcohol, and acid. Other sections discuss commercial wine disorders, and a more "advanced" chemistry of wine analysis. The author was one of the founding members of the British Amateur Winemakers.

68. Fornachon, J. C. M. **Studies on the Sherry Flor**. Adelaide, Australian Wine Board, 1953. 139p. illus. bibliog. LC A56-6493. out of print.

A classic research book, very technical, but one of the few that document the sherry flor. The author has also written *Bacterial Spoilage of Fortified Wines* for the Australian Wine Board (1943). Recent advances from the Davis Campus of the University of California indicate great success with an enforced and submerged flor.

69. Hough, James Shanks, D. E. Briggs, and R. Stevens. **Malting and Brewing Science**. London, Chapman and Hall, 1971; distr. New York, Barnes and Noble. 678p. illus. bibliog. £10.00. LC 75-860722.

A comprehensive, up-to-date account of the biological, biochemical, and chemical aspects of malting and brewing. It presents the scientific principles behind the selection of raw materials and their processing, including a description of equipment used. The details of practice are related not only to the scientific background but also to historical reasons and present economies. An international section described current practices and methods used in other countries.

70. Joslyn, Maynard A., and Maynard A. Amerine. **Dessert, Appetizer and Related Flavored Wines: The Technology of Their Production.** Berkeley, University of California, Division of Agricultural Sciences, 1964. 483p. illus. $7.50. LC 64-65053.

This document was prepared as an aid for the California wine industry in improving the stability of aperitif, dessert, and flavoured wines, and in increasing their acceptance and use by the consumer. Basic principles of winemaking are emphasized. Other subjects covered: choice of grapes, sensory judging, economies, winery design; origins of dessert and aperitif wines; flor sherries in California; vermouth; and wine disorders.

71. Ribereau-Gayon, Jean, and Emile Peynard. **Traité d'oenologie**. Paris, Librarie Polytechnique Béranger, 1966. 2 vols. illus. 270 Fr. F., set. LC 67-81369.

The French counterpart to Amerine, with good chapters on modern vinification methods and sensory evaluation.

72. Singleton, Vernon L., and Paul Esau. **Phenolic Substances in Grapes and Wines, and Their Significance**. New York, Academic Press, 1969. 282p. illus. (Advances in Food Research Supplement, 1). $14.50. LC 65-84257.

A complex, technical treatise dealing with the composition of grapes, oxidation and spoilage, "browning" and processing of wines, and consideration of the role of phenols as nutrients, pigments, and flavors of wine.

73. Underkofler, Leland, and Richard J. Hickey, eds. **Industrial Fermentations**. New York, Chemical Publishing Co., 1954. 2 vols. illus. $12.00 each. LC 54-7960.

Despite its age, this is still a basic book, written in clear language, with not too technical explanations. Volume One (564 pages) is divided into four parts. Part One describes the alcoholic fermentation process and its modifications (grains, molasses, wood, brewing, and the commercial production of table and dessert wines). Part Two is about the production of yeast. Part Three describes the butanol-acetone fermentations, while Part Four goes into the fermentative production of organic acids (fumaric, lactic, citric, acetic, etc.). Volume Two, of equal length, is more technical, dwelling on microbiological processes, the production of enzymes, vitamins and pharmaceuticals. The consumer, if interested in technical literature, would appreciate the first volume only. Chapters were written by 25 experts.

74. Winkler, Albert J. **General Viticulture**. Berkeley, University of California Press, 1962. 633p. illus. $12.50. LC 62-18717.

As a comprehensive compilation of contemporary practices in winemaking, this is a manual for vineyardists. Pruning is discussed, as well as climate and reasons for growth (or lack of it). The description of various wine grapes and table grapes is very important for the amateur winegrower.

GENERAL WORKS

This section is concerned primarily with those broad general works that encompass alcoholic beverages as a whole. To be noted especially are the profiles of the drinking public and the materials covering administration of government legislation and regulations. Encyclopedic materials are covered in the REFERENCE section above.

75. Acheson, Keith. **Revenue vs. Protection: The Discretionary Behaviour of the Liquor Control Board of Ontario**. Ottawa, Carelton University, 1972. 28p. charts, free.

This is the only liquor study available on the role of the state monopoly in determining what is available in the jurisdiction and what its price will be. Acheson looks into the mark-ups and concludes that they are tariffs (which it cannot legally do). Discretion is also a moral and ethical issue (what is available at what price). Applicable to those American states that have liquor control. Exceptionally readable.

76. Branch, S. N. **Special Report on Practices and Principles of Liquor Control in Some Canadian Provinces**. Halifax, N.S., Atlantic Provinces Economic Council, 1959. 57p. charts. out of print.

A descriptive study of prohibition in Canada and the resultant formation of Liquor Control Boards in each province. The practices, again descriptive, of four Eastern and four Western provinces are given, with no conclusions. Statistical appendices, and a short bibliography conclude the work.

77. Brewers Association of Canada. Alcoholic Beverage Study Committee. **Beer, Wine and Spirits; Beverage Differences and Public Policy in Canada**. Ottawa, 1973. 164p. tables. bibliog. free.

This report claims to be the first in the world to consider all the factors relevant to an appropriate public policy for alcoholic beverages. With analysis of data from 30 countries, this comprehensive review delved into the characteristics of beverages, their use in Canada and the world, the physiological effects, the problems raised by alcohol consumption, the encouragement of responsible drinking practices, a comparison of taxation and control practices, and economic analyses. As beverages of lower alcoholic strength are *least* harmful physically, there is an argument for progressive taxation per unit of alcohol. This study does such a superb job of resolving conflicts in the literature that we have felt it redundant to include the materials that the report is based upon.

78. British Columbia. Department of Education. Curriculum Development Branch. **The Knowledge and Serving of Alcholic Beverages**. Burnaby, B.C., 1972– . Various pagings. photos, diagrams. free.

This is more or less a condensed version of Haszionic's book (see entry 112), suitable for instruction in a junior community college. Everything is simplified and reduced to its basic elements. Copious charts, descriptions, and definitions. Very applicable to home use, especially for parties. This appears to be the best instruction book on the market thus far for elemental barkeeping, preparation of mixed drinks, equipment, recipes for beginners, wine waiters.

79. Butler, Frank H. **Wine and Wineland of the World, With Some Account of Places Visited**. New York, Finch, 1971. 271p. illus. $15.00.

This is a reprint of the 1926 edition, originally published in London by T. Fisher Unwin. It opens with a discussion of religious and health reasons for consuming wines, thereby getting off on the right foot. Most of the 55 photos are the author's. World traveller Butler covered Portugal, Spain, France, Italy, Algeria, Morocco, Russia, South Africa, Australia, Argentina, Chile, Canada and even far-flung non-wine-producing areas (at least in quantity) such as Kashmir, Norway, and the South Sea Islands. Covered are brandy, cognac, whisky, Irish whiskey, Bacardi rums; English ale, perry, gin, Chartreuse, Sabri; drinking songs, fashions in drinking wines, and even balloon voyages.

80. Cahalan, Don, Ira H. Cisin, and Helen M. Crossley. **American Drinking Practices: A National Survey of the Behaviour and Attitudes Related To Alcoholic Beverages**. Princeton, N.J., Rutgers Center of Alcohol; distr. by College and University Press, 1969. 260p. Charts. bibliog. (Rutgers Center of Alcohol, 6). $9.50. LC 70-626701.

There were two main purposes behind this survey (originally done for the Social Research Group at George Washington University in 1967): to analyze the demographic and sociological correlates of levels of drinking (e.g., amount, socioeconomic status, rate, national origin, religion, sex, age, type of alcoholic beverage, etc.), and to study the range of drinking practices (*why* specific beverages were consumed and when). The authors used questionnaires, surveys, and interviews–in all, more than 3,000 persons were queried. Their recreational activities were explored. Appendices include the questionnaire and survey questions, tables of computations and so forth. Other titles in this Rutgers series includes No. 3, *Alcohol in Italian Culture (Food and Wine)*, and No. 5, *Drinking in French Culture.*

81. Cavan, Sherri. **Liquor License: An Ethnology of Bar Behavior.** Chicago. Aldine Publishing Co., 1966. 246p. $7.50. LC 66-15199.

Generally, this work purports to study the face-to-face interaction of people in their drinking places. In four kinds of bars (home territory, convenience, sexual meat markets, and night spots), the rules of the game (or the rituals and ceremonies) are examined; who talks to whom; how pick-ups are arranged; and how to buy drinks and "rounds" for other people. Recognizable behavior traits emerge, no matter where the drinker is, be it saloon, lounge, night club, roadhouse, cabaret, beer garden, bistro, or pub. Fascinating reading. For historical items, see the separate section below (p.120).

82. Chafetz, Morris E. **Liquor: The Servant of Man**. Boston, Little, Brown & Co., 1965. 236p. $5.95. LC 65-15238.

The author, a psychiatrist, stresses the positive aspects of liquor consumption: the pleasures of imbibing. Medical aspects are examined, and a brief history of man's involvement with alcohol is given. Of course, sexual mores are mentioned in the section detailing liquor use around the world.

83. Durkan, Andrew. **Vendange: A Story of Wine and Other Drinks**. New York, Drake, 1972. 327p. illus. maps. $10.00. LC 75-868562.

This is a textbook for advanced students of hotel administration. Eighty pages (one quarter of the total) deal generally with beer and spirits. The author was formerly an inspector of hotels in Ireland, and he is now a lecturer in the food industry. "Vendange," which means the gathering of grapes, is freely translated to mean the gathering of knowledge about alcoholic beverages. The coverage is worldwide, but it is out-of-date is some areas (e.g., New Zealand).

84. International Union of United Brewery, Flour, Cereal, Soft Drink and Distillery Workers of America. **Union With a Heart: 75 Years of a Great Union (1886-1961)**. Cincinnati, Ohio, 1961. 32p. illus. free.

The value of this pamphlet lies in the many historical pictures from the Union files—pictures unavailable elsewhere.

85. Joint Committee of the States to Study Alcoholic Beverage Laws. **Alcoholic Beverage Control**. 2nd ed. rev and enl. Washington, D.C., 1960. 114p., tables. free. LC 60-64195.

"Of particular interest to public officials, to students of government and to all persons seeking factual information and authoritative guidance in this field of governmental regulation of the only industry destroyed and recalled to life by amendments to the Constitution of the United States." This is the history of government control, including Prohibition, Repeal, and the Local Option. The book examines the whole complex of the administrative structure (rulemaking, regulations, internal management, judicial review), and the concept and development of both licensing and enforcement. There are extensive tables of facts and figures.

86. Joint Committee of the States to Study Alcoholic Beverage Laws. **Uniform Standards for Advertising of Alcoholic Beverages in Newspapers and Magazines**. Washington, D.C., 1963. 114p., illus, biblio. free. LC 63-64565.

This brief history of alcoholic beverage advertising delineates the responsibilities of the two levels of government, and analyzes the requirements and prohibitions of the states and of the federal government. This exhaustive study, well illustrated with old advertisements, concludes that the confused maze of legislation and regulation makes it difficult to enforce and act against improper and irresponsible advertising. Advertising in interstate media remains unaffected by state control, and there appear to be special privileges for malt beverages. This readable book concludes with statistical information and summaries.

87. Lewis, Jack, pseud. **Official Liquor Buyer's Guide**. Los Angeles, Holloway House, 1969. 213p. $1.50pa.

Information on how to buy good liquor at the best price, and how to tell good liquor from bad. Most of the book is descriptive of product, including what is involved in the manufacture of liquor (with some trade "inside" information), some 10 pages of recipes, a 20-page glossary, and 30 pages of best-buy samples for Pennsylvania, California, Illinois, Colorado, New York, and Hawaii. But the expanded "best buys" charts and those of other states are available only by subscription for $2.00 from: P.O. Box 25929, Los Angeles, California 90025. This is an *annual* subscription and updating service.

88. Licensed Beverage Industries, Inc. **The Alcohol Beverage Industry: Social and Economic Progress**. Washington, D.C. 1973. 48p. charts. free.

This annual report is on the industry's direct and indirect contributions to the national economy since Repeal. Statistics include the state-by-state analyses of industrial employment, payroll, and taxes.

89. Mahoney, John. **Wines and Spirits: Labelling Requirements**. 2nd ed. London, Wines and Spirits Publications, 1972. 60p. $3.00pa.

Since Britain joined the EEC in 1972, this book, meant for the English trade, had to be rewritten. In its present form, the EEC regulations are covered. Mainly of value for those in the British wine trade who are concerned with bottling and labelling, the work covers all the laws with illustrative examples for wording and placement on the label. A valuable section of the book consists of a wide selection of specimen labels, and also a section dealing with offenses, defenses, and penalties.

90. Marrison, L. W. **Wines and Spirits**. 3rd ed. Baltimore, Penguin Books, 1973. 335p. illus. maps. photos. $1.95pa.

Marrison's guide to wines and spirits was first published in 1957 and had undergone several revisions. Part One covers an introduction to the quality of wines, wine countries, history of wine, four chapters on the making of wine and kinds of wine, vineyards of the world, and how to drink wine. Part Two covers spirits: brandy, whisky (including whiskey), rum, gin, other spirits.

Additional information at the end covers wine-growing in England, table of measures, maps, a classified bibliography, and index. A reading of Marrison's section on pests and diseases proves that more information is packed into less space than in any other wine guide encountered. He is exceptionally good on wine chemistry. This book is packed with information, and he leaps right into his points with few preliminaries. Thus, he is a little difficult to read sometimes, and his book is more for study than for perusal.

91. Mendelsohn, Oscar. **The Earnest Drinker: A Short and Simple Account of Alcoholic Beverages (With a Glossary) for Curious Drinkers**. New York, Macmillan, 1950. 241p. out of print. LC A52-9344.

Originally published in Sydney, Australia in 1946 under the title "The Earnest Drinker's Digest." The author deliberates on the fallacies of alcohol, how alcohol is formed, its relationship to medicine and to health, and the various processes involved in brewing and distillation. The usual spirits are covered (as well as cider) with some notes on their coloring and strengths. Storage and service of wines are discussed. In fact, here there is a lot of varied, accurate, and well-classified information. Both the preface and a special 20-page section on French, German, and South African wines were written by T. A. Layton.

92. Mew, James, and John Ashton. **Drinks of the World**. Ann Arbor, Mich., Gryphon Books, 1971. 366p. illus. LC 70-78207.

This reprint of the 1892 London edition contains "one hundred illustrations." At the time, it was intended for the general but informed reader. In coverage, the last hundred pages deal with non-alcoholic drinks such as tea, coffee, and chocolate. Drinks of antiquity lead off, with discussion on the Egyptian method of making wine, beer vessels and goblets, the Greek spiced wines, use of resin, amphorae (illustrated) the meads and ales of Scandinavia, and so on up to the end of the nineteenth century. Short histories are given on wine-growing abroad, notably in Africa, Australia, and the United States. Large sections are devoted to ciders, brandies, gins, whiskies, rums and liqueurs. (The chapter on liqueurs describes many that are no longer made.) Some recipes are given for German and French liqueurs, many still workable. "American drinks" is really a chapter on cocktails (the term had not come into general usage then), and a few passé drinks are covered: shrubs, cobblers, sangarees, etc. Many recipes are here, for example, "A Yard of Flannel," "Bottled Velvet," "Stone Fence," etc. The last chapters, on beer, are international, with coverage extending to China and Borneo. Much like a dictionary on historical principles, Mews and Ashton present miniature histories for each drink or liqueur. "Toddy," for instance, was originally a Hindustani word for a tree sap. While there is a wealth of material here, and while it is superbly organized, the book is crammed haphazardly with liberal doses of source materials and quotes, and the overall readability is thus reduced.

93. Sheperd, Cecil W. **Wines, Spirits and Liqueurs**. New York, Abelard-Schuman, 1959. 160p. illus. $4.50. LC 57-12316.

This book, published a year earlier in England, is really a basic text for the ordinary person. All countries are covered, and spirits include rum, gin, brandy, and whisky. Anecdotes predominate. The appendix details a selection of table and dessert wines.

94. Waugh, Alec. **Wines and Spirits**. New York, Time-Life Books, 1968. 208p. illus. maps. (Foods of the World). $7.95. LC 68-55300.

This is very much a continuation of his *In Praise of Wine* (see entry 458). His personal style, almost like a diary, recounts the brief history of wines and spirits. Still, it reads like any other basic introductory text. There are superb photographs by Arie de Zanger, and the book is worth buying for these alone: they are among the best to be found, and they are particularly valuable for their illustration of techniques. Appendix material covers vintage years, vineyards of France, glasses, bottles, labels, wine and food, wine-tasting parties, and starting a wine cellar (with suggestions for three levels of budget spending). This is the only book in the Time-Life series *without* recipes in the text. The *Recipe Book* (part of the total price above) is available separately to libraries for $1.45. It is spiral bound, with washable covers, and a fourth of its 96 pages are devoted to hors d'oeuvres and canapes. There are explanatory recipes for distilled spirits, wines, punches, and hot drinks, plus a concluding section on a basic equipment list.

95. **Wines and Spirits of the World**, Ed. by Alec H. Gold. 2nd ed. fully revised. Chicago, Follett Pub. Co., 1972. 753p. illus. $19.95. LC 71-177450.

This posh British book is geared to hotels and restaurants. Each chapter was written by a specialist (e.g., "Germany" by Hallgarten, "U.S.A." by Gold). Part One, Wines, occupies 550 pages. The arrangement is geographical, so one must know a wine's provenance before using the book. France, of course, gets half of the space alloted for Part One, followed by Germany. In the first British edition (1968) North America was represented by two pages for United States wines. The second edition is 45 pages longer, and all of this coverage goes into Canada and the United States. Sherry, port, madeira and vermouth are also in this first section. Part Two is concerned with spirits–brandy, whiskies, vodka, gin, rum, liqueurs, cocktails, cider, and beer. The section on apertifs appears to be weak, and in the liqueurs section no differences are noted among similar types (e.g., Tia Maria vs. Kahlua vs.

Zarankaffee). There are special sections on serving, storing, tasting, drinking glasses (British), and label information. Generally, the allocation of space for country of origin is equal to the types and quantities of alcohol available in Britain. The style and nature of the book are far too salutary when describing sweet wines, and there is a lot of white space and huge print on each page. The 35 two-color maps give the basic details, but the two dozen color plates and other photographs complement both the text and the 200 line drawings.

Chapter 2

WINES

GENERAL WORKS

Works in this section deal with the general subject of wines—what they are, how they are consumed, and why some are better than others. It is in this area that gross duplication can be found, and it was difficult to come up with 40 or so books that really had something different to say. The new material in many of these books could be reduced to a few pages, yet it is this unique material or method of expression that saves them from oblivion. Nevertheless, there is an appalling lack of data that actually contribute to furthering the development of wine literature.

First of all, the books include very few tastings and tasting tips. Much of this material is found in wine periodicals, but little in permanent hard covers, where one day it might be of fine historical value or might assist in tracing the changes in one vineyard's vintage years as it reaches its peak. See the separate section on tastings (starting with entry 481), and note the scattered references to tastings within the annotations in this bibliography.

Second, the examination and corporate histories of merchants, brokers, growers, and other persons in the wine trade are missing, primarily because most of the wine makers produce books about their products, not about themselves. No wine merchant has yet written his memoirs as they affect the trade. No outsider has ever investigated the practices of the wine trade. It is a very closed and tight operation.

Third, as is evidenced by the paucity of entries under a specific country, more detail on vineyards and the small areas is needed, along with vintage charts or explanations for individual vineyards or regions. More details are needed about the wines of various vintages—production, bottling, storage, and so forth.

Fourth, there should be more description of wines not yet available in the United States, for a number of reasons: many of these are country wines that will soon be available as the classified growths price themselves off the market; many are already being imported anyway just as a matter of course, and there is a long wait between book revisions (if any); travellers would like to read about these wines when they return home; and finally, such descriptions would provide variety in the wine literature—something new for everybody to learn. British books have some of this information as their wine markets are different from those in North America.

Many books appear to be part of a cohesive series, apparently published in that way in the hope that readers would buy all volumes of the series, piggy-back fashion. The first series, which dates from the mid-thirties, was Constable's Wine Library. Because the books were descriptive only, being practical handbooks on wine and appreciation, and because some are almost 40 years old and also out of print, they are not described in this book. The series included *Sherry* (H. Warner Allen), *Champagne* (André Simon), *Claret and the White Wines of Bordeaux* (Maurice Healy), *Madeira: Wine, Cakes, and Sauces* (André Simon and Elizabeth Craig), *Burgundy* (Stephen Gwynn), *Port* (André Simon), *Wine in the Kitchen* (Elizabeth Craig), and *Hochs and Moselles* (Hugh Budd). The latter is included in the German section because it is a definitive study. During the early 1960s, McGraw-Hill put out a series of wine books in conjunction with a number of British publishers (especially George Rainbird); when released in the United States, the books were uniform in size and method. The other American series of note would be Hastings House's "Drinking for Pleasure Series." At present, the current publisher series of wine books is Faber and Faber's excellent "Library of Wine" (Britain), a superb series that includes Julian Jeffs' *Sherry* and *Wines of Europe*.

96. Adams, Leon D. **The Commonsense Book of Wine**. Rev. ed. New York, D. McKay Co., 1964. 178p. (Tartan Books, 17). $1.45pa. LC 65-2146.

Although written in a haphazard fashion, this book, first published in 1958, covers a lot of ground: how wines taste, their chemistry, costs, restaurants, explanation of labels, hobbies, medical aspects, "exposing" wine snobbery, and why people drink wine. Coverage is only of wines that are available in the United States. There are charts of color, flavor, and alcoholic content.

97. Bain, George. **Champagne Is for Breakfast**. Toronto, New Press, 1972. 277p. illus. maps. $6.95.

A former wine columnist, Mr. Bain has gathered his best notes into a weird book made more difficult because of his digressing style (with cute chapter headings). Some duplication with other books (mainly in giving unorganized data). Yet there are plentiful anecdotes of the wine trade, critical analyses of the monopolistic liquor commissions, and about two dozen top-notch recipes for using wine in cooking (e.g., Daube Avignonnaise). Some travel notes and personal accounts are given, along with an explanation of the 1971 German Wine Laws. The first 200 pages are devoted to France. Definitely not a book for the novice.

98. Beck, Frederick K. **The Fred Beck Wine Book: Fred Beck's Gay, Clearcut Explanation of the Wines of the World, with Special Recognition of the Wines of California**. New York, Hill and Wang, 1964. 242p. illus. maps. plates. $4.95. LC 64-24829.

This is generally a book that is fun, full of anecdotes and gags. It provides a painless way to get to know wines, but there is little else to recommend it, except for a long section on judging wines. The subtitle tells all.

99. Bespaloff, Alexis. **The First Book of Wine**. New York, World Publishing Co., 1972. 232p. illus. $7.95; $1.25pa.(available from Signet Books). LC 72-183152.

Originally published in paperback in 1971. A comparison of the hardbound version with the original reveals that it is a straight photographic enlargement (huge print, broken type, etc.). This book is ordinary in comparison with others of like structure, but the impact of the paperback was very strong. First of all, Bespaloff is widely published in his field (mostly periodical literature). Secondly, the book was promoted as a first introduction to wine. And thirdly, it was the first wine book ever published in paperback that had a wide, national distribution. The usual information is contained here: general information; descriptions of wines and countries (yet only 13 pages on the United States); material on champagne, fortified wines, and cognac; and a short pronouncing guide. Consumer information includes wine cellars and selection of low-priced wines (prices are out of date already), vintages, and how to serve wines. All in all, relatively good value for $1.25—but not for $7.95.

100. California Wine Advisory Board. **The Story of Wine and Its Uses: A Non-Technical Guide to Wine**. 8th ed. San Francisco, 1970. 49p. illus. free.

Although the emphasis is on California wines, this guide provides compact and useful information. Skillful writing produces material on wine types, how wine is grown, wine quality, the history of wine, and the modern industry in California today. Well illustrated, with appropriate black and white photographs. This pamphlet is used as a text for a free correspondence course run by the Wine Institute.

101. Churchill, Creighton. **The Great Wine Rivers of Europe**. New York, Macmillan, 1971. 256p. illus. $9.95. LC 79-158162.

This is a motorist's guide, similar to Wildman's efforts for France (see entry 152); it covers the major wine districts found in the Moselle Valley, Rhine, Rhone, Loire, and Gironde. Detailed description is given for the country, vineyards, grapes, and characteristics of the wine produced in each region. Suggestions for travel are made (roads, hotel, restaurants, winetasting) by both Churchill and his wife. One drawback: no maps.

102. Churchill, Creighton. **The World of Wines**. New York, Macmillan, 1963. 271p. map. bibliog. $6.95. LC 64-21761.

Some of these essays first appeared in *Harper's* and *Gourmet*. Even so, the material is pretty elementary: what wine is, France (one-third of the book), Germany, Italy, Portugal, and Spain, followed by sections on champagne, fortified wines, and brandies. The book concludes with advice on wine selection, storage, and serving. Each chapter is summarized and followed by a list of principal wines. Vintages are evaluated, but country wines are stressed.

103. Clarke, Nick. **Bluff Your Way in Wines**. New York, Crown, 1971. 63p. illus. maps. tables. (The Bluffers Guides). $1.00. LC 68-104175.

Originally published in England four years earlier, this straight reprint is an amusing little book, dedicated to the lady who described an inoffensive Moselle as being "tender, without being mawkish." The inclusion of inside jokes makes this book a jewel for the professional, the advanced amateur, or the wine snob, with added tips on how to combat such snobs. Plentiful information is presented, in almost point-by-point form. The book deals with: selection of wine, tasting, merging of food and wine, choosing a wine in a restaurant, and dominating a wine steward. Despite such obvious statements as "wine glasses should be water tight," it does include facts not commonly known, such as a description of a white Beaujolais. Coverage is worldwide, with small detail on sherry and port. The book closes with a "Totally Useless Glossary."

104. Don, R. S. **Teach Yourself Wine**. New York, Dover, 1971. 202p. illus. forms. maps. $2.50pa. LC 70-355671.

This introductory text, with concise information and rudimentary maps, contains a buyer's guide that is geared primarily to the British market (a reflection of its original 1968 English edition) and that is dated by the inclusion of prices. The first six chapters describe the winemaking process. The balance of the book details wine drinking, buying, and cellarage, and choosing and serving wines. The end papers have a chart of wines and matching foods.

105. Dorozynski, Alexander, and Bibiane Bell. **The Wine Book: Wine and Wine Making around the World**. New York, Golden Press, 1969. 310p. illus. $12.95. LC 69-19127.

Originally published in Paris and Verona in 1968, the historical approach used here delineates *how* wine is vinified around the world. There are major sections on burgundy, champagne, sherry, port, tokay, and chianti. Some out-of-the-way places include the USSR, North Africa, the Near East, and Latin America. At the conclusion, there are 50 typical recipes, mostly for the main course. The 260 illustrations (half in color) include historical wine labels, presses, harvesting, and wine lists from long-gone restaurants. Wine labels (modern) are scattered among the end papers.

106. Edita Lausanne. **The Great Book of Wine**. New York, World, 1970. 459p. illus. maps. $50.00. LC 77-124428.

A coffeetable book that describes 6,500 different wines, but with only 54 labels reproduced in color (grouped by country of origin) and explained. The gorgeous 44 maps and tipped-in illustrations, however, are not enough to carry the defects in the rest of the book–namely, the price, and the lack of an overall index (only wine names are indexed). Other surveys written by wine specialists (one to a country) are better; Simon (entry 130) and Johnson (entry 115) say more, and their books are cheaper. The usual collective data on storage, serving, and tasting can also be found elsewhere. Still, a very impressive decorative book that is a work of art.

107. Fluchère, Henri. **Wines**. New York, Golden Press, 1973. 160p. illus. maps. bibliog. $1.95pa. LC 72-85934.

This is an unqualifiedly "superb" book. It is one of the Golden Handbook series that fits easily into pocket or purse. While it is very general, the book is also comprehensive, giving adequate coverage to winemaking (good notes and pictures on grape varieties), the various country wines, including even Russia and Japan, and descriptive material on California wineries, how to read labels, and so forth–all in the first 100 pages alone. The balance of the book is devoted to enjoying wine: shopping, storing, serving, wine-tasting parties, wining and dining out, wine and weddings (e.g., champagnes), wine and food (including what usually goes together), how to visit wineries and museums (worldwide, including Spain and Siena, Italy), and a concluding glossary. The illustrations by the author, an accredited heraldic artist, appear on every page, and each is in color (the book is published on clay-based glossy paper). Fluchère's work is obviously intended for today's ordinary urban consumer. Lots of labels are shown, with data on how to read them; there are seven excellent pointers on what to look for when shopping for wines; there is an illustrated description of an apartment wine "cellar" plus a detailed description of how to construct wine racks. The food chapter includes a "first"–what wine goes with hash (rosé) and hot dogs (dry white)! Quite possibly this is the best book currently available for the wine novice.

108. Food and Agriculture Organization of the United Nations. **The World Wine and Vine Products Economy: A Study of Trends and Problems.** Rome, 1969. 50p. illus. (FAO Commodity Bulletin Series, 43). $2.00. LC 74-579893.

This book is the result of three 1967 studies dealing with wine, table grapes, and dried vine fruits (raisins). The wine section (24 pages) gives a brief account of trends of world production and trade in recent years. The problems and difficulties facing each country are examined, along with treaties and potential markets. Despite this activity, international trade in wine is less than 10 percent of total production.

109. General Agreement on Tariffs and Trade. **World Trade and Prospects for Ordinary Wine**. Geneva, 1966. 92p. charts. $1.00. (Distr. by Unipub).

Leading into the 1969 work, *The Market for Wine in Ten Western European Countries*, this publication explores the nature of international trade for the less common wines, finding a healthy exchange for the countries bordering France and Italy. No conclusions were drawn. Good supplementary material on charts, figures, and so forth. (See also entry 113.)

110. Grants of St. James. **A Gateway to Wines**. London, 1964. 76p. illus. free.

This delightful little hardbound book is well illustrated and simple, yet lavishly designed. After a description of wine-producing areas, there is a short guide (mainly to its own imported wines), then the obligatory handful of recipes and a glossary.

111. Hasler, G. F. **Wine Service in the Restaurant**. London, Wine and Spirit Publications, 1972. 84p. £1.00.

The author, a founding member of the Guild of Sommeliers, has written this book primarily for students and trainees in the hotel and restaurant trade. It covers the serving of wine in cocktail lounges and restaurants, catering for restaurant parties and banquets, and other subjects that may contribute to the success of the dinner party at home as well. Strangely enough, there is nothing on that decade-old phenomenon–wine bars.

112. Haszonics, Joseph T., and Stuart Barratt. **Wine Merchandising**. New York, Ahrens Book Co., 1963. 214p. illus. maps. facsims. $8.00. LC 63-15806.

Haszonics, a restaurant magazine editor, and Barratt, a Las Vegas sommelier, have written the classic book whose "purpose is to help the restaurant sell wine," in order to maximize the beverage dollar. Facets of service examined include: instruction in the wine cellar, wine list, stemware, wine labels; pointers on employee training, merchandising display, and the role of the wine steward; notes on winetastings, chemistry and history; and an invaluable pronunciation guide. Wine labels are reproduced, with a clear explanation of what each label means. Individual chapters discuss wine from Europe and the United States from a merchandising perspective. This superb book also is really instructive for the ordinary student of wines.

113. International Trade Centre. **The Market for Wine in Ten Western European Countries**. Geneva, 1969. 402p. tables. $1.00. LC 77-860728.

Under the joint auspices of GATT and UNCTAD, this worthwhile publication presents an analysis of regulations in the EEC and the former EFTA for its member countries of the United Kingdom, Austria, France, Germany, Belgium, Sweden, Denmark, Norway, Switzerland, Netherlands. Each country is examined as to: production, consumption, external trade, import restrictions, legislation applicable to wines, consumption patterns, trade practices, trade promotion, distribution channels, prices and margins, and transportation. (See also entry 109.)

114. Johnson, Hugh, ed. **The Pan Book of Wine**. 2nd ed. London, Pan, 1964. 176p. illus. bibliog. $1.00.

In a number of articles reprinted from *House and Garden* (1954-59, and 1964), 27 experts describe and review wines of the world, with information on how to buy, to decant, to taste and to enjoy. Contributors include such well-known names as Raymond Postgate, André Simon, Evelyn Waugh, Harry Waugh, and Baron Philippe de Rothschild.

115. Johnson, Hugh. **Wine**. New York, Simon and Schuster, 1966. 264p. maps (col.). plates (col.). $12.00. LC 67-8822.

Also available from England as a 347-page paperback (Sphere, 1970) for $2.95. This is perhaps the best all-round introduction to wine. This very readable guide to wines and vineyards of the world is really an encyclopedia written in narrative form; it complements, of course, Johnson's *Atlas* (see entry 19). The traditional pattern of description is followed: how wines are made and served, their history, wine selection, matching wines with food—subdivided by "Aperitifs," "White and Red Table Wine," and "After Dinner Wines."

116. Kressman, Edward. **The Wonder of Wine**. New York, Hastings House, 1968. 227p. $6.95. LC 68-20253.

As a manager of the family wine farm, Kressman has access to inside information that is presented in this good book of memoirs, a sort of "my life in wine" adventure. Some of the material was previously published in *Wine* magazine, but the personal accounts were not. There is good explanatory material for the layman on the first steps in fermentation and racking. He takes an extensive approach to the wine bottle and to its pedigree, which includes the label, shape, cork, date, grape variety, price, decanting, etc.

117. Layton, T. A. **Modern Wines**. London, Heineman, 1964. 190p. £1.50.

As a very prolific author, Mr. Layton has covered the whole world of wines through time and space. This time out he reviews the lesser-known wines of some 18 different countries (e.g., Chinon and Bourgueil from the Loire), including Alsace, Wachau, Italian Piedmont, the Egri Bikaver, plus a great deal of material on Latin America, Algeria, Cyprus, Israel and Rumania, among others. The appendixes: the state of viticulture in the world, and some international viticulture statistics.

118. Leedom, William S. **The Vintage Wine Book**. New York, Vintage Books, 1963. 264p. illus. $1.95pa. LC 62-12737.

Another original work that concentrates on France, Germany, and California, with cogent advice on buying and selecting wines.

119. Marrison, L. W. **Wines for Everyone**. New York, St. Martin's, 1971. 212p. illus. $6.95. LC 77-134840.

Perhaps a better book would be his *Wines and Spirits* (Penguin, 1973; see entry 90). Yet this introduction will do admirably well in the field with the couple of hundred other books. The author's humorous touches lend levity to the proceedings.

120. Massee, William Edman. **Massee's Wine Handbook**. Rev. ed. Garden City, N.Y., Doubleday, 1971. 216p. illus. $5.95. LC 73-131095.

Massee's handbook, first published in 1961, offers some help to the newcomer to wine buying. The book covers kinds of wine, how to read labels, red wines, white wines, pink wines, champagne, aromatic wines, and fortified wines. There is no attempt to be comprehensive; only major wines are introduced and described. The purpose of the handbook is to provide some assistance to the bewildered buyer. The 56 drawings of wine labels help to make the text more like a trip to the neighborhood liquor store, although such illustrations are not needed when the actual wine label could be had for proper reproduction. Most of the book is written in virtual chart form. Price ranges and vintage charts are updated to 1970, but this updating needs to be undertaken again for modern times.

121. Misch, Robert T. **Quick Guide to Wine: A Compact Primer**. Garden City, N.Y., Doubleday, 1966. 98p. illus. $3.50. LC 66-20948.

Bonvivant Misch, who writes a syndicated column about wine and food, here attempts to class wine by categories and colors. The origins and qualities are carefully explained, and much advice is given on serving, storing, and food, with ample charts detailing the appropriate "what goes with what." For the absolute novice.

122. Postgate, Raymond. **An Alphabet of Choosing and Serving Wine**. New York, Wehman, 1959. 94p. $1.00pa.

123. Postgate, Raymond. **The Home Wine Cellar**. New York, Wehman, 1964. 92p. $1.00pa.

124. Postgate, Raymond. **The Plain Man's Guide to Wine**. Rev. ed. London, Michael Joseph, 1970. 164p. illus. map. £1.50. LC 71-517408.

Postgate is always easy to read, and he is very reliable when it comes to gathering facts and balancing judgment. The first two books are very slim, but they deal with specific items such as choosing a wine for home use, and matching it with the "proper" food. Cellarage is discussed in the second book–how to build, how to store, rate of withdrawal, and so forth. The *Guide*, now in its third edition, is a basic text geared to the British market and descriptive of wines around the world. It is particularly valuable in view of Postgate's previous track record of enviable accuracy.

125. Pratt, James Norwood. **The Wine Bibbers Bible**. San Francisco, 101 Productions, 1971. 191p. illus. $6.95; $3.95pa. LC 79-177349.

A glamorous "production" (not "publication") for those who want an artsy-craftsy type book to explain what they could already know. If this is the only way to get a reaction from people, then so be it. Information, while generous, is duplicated from other sources, but the eclectic-minded may appreciate the translated Baudelaire essay or the Raffetto drawings. For the wine snob.

126. Rainbird, George M. **The Subtle Alchemist: A Book of Wine**. London, Michael Joseph, 1973. 206p. illus. bibliog. $10.95.

This was originally published as *The Pocket Book of Wine*, which was itself derived from Rainbird's *The Wine Handbook* (Hawthorn Books, 1964). This time around, the practical and realistic Rainbird is assisted by Ronald Searles's hilarious cartoons. Also, there is much more information on country wines, a wise addition in view of today's escalating prices. So Searles and country wines account for the 50-page discrepancy between the former book and this present one.

127. Sichel, Allan. **The Penguin Book of Wines**. 2nd ed. with revisions by Peter A. Sichel. Baltimore, Penguin Books, 1971. 304p. bibliog. maps. $1.75pa.

The original edition was published in 1965. This revision updates prices, new wine laws, and vintage notes, but the main text remains the same. Its purpose is to serve as an introduction to the whole subject of wine, and it successfully achieves this goal. The four parts are well organized; they cover general information (tasting, wine names, storing wine); vineyards, soil, winemaking, the grower, shipper and merchant; an introduction to European wines by country, followed by short descriptions of non-European wines; a vintage list, a well-selected bibliography (briefly annotated), glossary, nine maps, and an index. While this is an admirable, inexpensive guide for the novice, it must be emphasized that the information is aimed at British readers. The maps at the back of the book are too far away from the related pages.

128. Simon, André L. **The Commonsense of Wine**. Cleveland, World Publishing Co.; distr. by Pyramid, 1966. 192p. illus. $3.95. LC 65-28630.

This short book answers 80 questions in a question-and-answer format. Answers are easy to find, if you know the right questions.

129. Simon, André L. **A Wine Primer**. Rev. ed. London, Michael Joseph, 1970. 167p. illus. maps. £1.50. LC 77-496797.

Another basic, commonsense introduction that does not talk down to the reader. This is the third edition (the other two appeared in 1946 and 1964), and it is the only edition that is indexed.

130. Simon, André L., ed. **Wines of the World**. Rev. ed. New York, McGraw-Hill, 1972. 719p. illus. maps. bibliog. $20.00. LC 67-14853.

This is a well-illustrated collection of monographs, but it has not been sufficiently revised since the 1967 edition. The sections by Simon himself have not been revised (he died in 1970). The brief bibliography lists no books published after 1965. Still, this compendium of the wines of the world was written by specialists such as S. F. Hallgarten (Germany), Simon (France), Cyril Ray (Italy), George Rainbird (sherry) and H. Warner Allen. The information is compact, and there is a glossary. In a way, this seems to be a sampler, since many of the sections were published by McGraw-Hill as separates, with more illustrative matter and substantial statistical appendixes thrown in.

131. Street, Julian L. **Wines: Their Selection, Care and Service**. 3rd ed., rev. and ed. by A. I. M. S. Street. New York, Knopf, 1961. 243p. illus. $5.95. LC 60-53486.

The dean of American wine writers until he died in 1947, Mr. Street has written an account that emphasizes French and German wines, ports and sherries. Supplementary material includes: a chart of vintage years, observations on harmonies between certain wines and certain foods, wine glasses, wine cradles, corkscrews, and so forth.

132. Torbert, Harold C., and Frances B. Torbert. **The Book of Wine**. Los Angeles, Nash, 1972. 401p. illus. maps. $10.00. LC 73-167522.

All the major wine-producing areas in the world are covered, and the authors concentrate on wines of flavor and integrity by progressing from the rich Burgundy, through Bordeaux, the rest of France, Germany, etc., right on down to *vins ordinaires*. Much space is devoted to German wines, especially to those that are not very common in the United States. Miscellaneous material includes short notes on aperitifs and after-dinner drinks, problems of wine in restaurants, and how old wine should be when it is consumed. Not much is said on American wines. The style of writing is very readable and personal, sometimes to the excess of condescension. There are 20 pages of extensive tasting notes.

EUROPE

GENERAL

In this "General" section are *single* books on the wines of individual countries. When there are more than one book on a country, these are entered under that country's name.

133. Duttweiler, Georges, ed. **Les vins suisses**. Genève, Editions Générales, 1968. 311p. illus. bibliog. 46.50 Fr. Swiss. LC 70-371118.

These 16 chapters in French run through general material such as winemaking, tasting, types of wine, wine cellars, eaux-di-vie, wine and food, wine service, and doctors and wine (health aspects). More specific sections deal with producers (by canton) and regional gastronomic specialties, institutions and associations in Switzerland, legislation on wines and protection of the consumer, plus the role of wine and grapes in the economy (with relevant statistics). Other chapters relate a witty history of the wine, plus proverbs. There are lists of classical wine types, by canton and subdivisions within.

134. Gunyon, R. E. H. **The Wines of Central and South-Eastern Europe**. New York, Hippocrene Books, 1972. 132p. illus. $6.95. LC 76-880515.

This book rectifies the previous lack of information in English on good sound country wines from these neglected areas. The scarcity of premium wines at a low price will increase the day-to-day need for these ordinary products. This slim book does an admirable job (from Prague to Crete and Vienna and across to Bucharest) in covering Hungary, Yugoslavia, Bulgaria, and Russia. Much material on Russia, unfortunately, is available only in the Russian language.

135. Halasz, Zoltan. **Hungarian Wine through the Ages**. Budapest, Corvina Press; distr. by International Publications Service, 1971 (c.1962). 185p. illus. bibliog. $4.25. LC 63-729.

This is an historical text, with an extensive Hungarian bibliography but no index. Each district is discussed: Somló, Mór, Szekszard, Egri, Balaton, and Tokay (among the major areas). There are chapters on food and wine, folklore, taverns, inns, and wine drinking, plus information on wine exports and the role played by Momipex (the state wine monopoly). The concluding section describes the scientific basis of the wine production in Hungary today. The illustrations are simple but humorous water colors. Naturally, Egri Bikaver is the pride of Hungary, and there are many pages devoted to its delights. But the book does not tell why quantities of wine are so uneven in the bottle: they are still filled by hand with rubber hoses.

136. Hornickel, Ernst. **The Great Wines of Europe**. New York, Putnam, 1965. 229p. illus.(part col.). maps. $12.50. LC 65-19759 rev.

The third German edition of this book was published in 1963. The basics are outlined, and wine peerage is discussed, with much detail on the great wines and vineyards of France, Germany, Austria, Switzerland, Italy, Spain, Portugal, and Hungary. Each country has its own listings of best vintage years. Special chapters describe champagne and the German Sekt, port, and sherry. The qualities and characteristics of various grape varieties are covered, and the appendix gives a wine cellar listing. Unfortunately, this is a difficult and turgid translation.

137. Jeffs, Julian. **The Wines of Europe**. New York, Taplinger, 1971. 524p. maps. $14.95. LC 76-153673.

The Wines of Europe was first published in Great Britain in 1970 and was written by a former editor of *Wine and Food* and author of *Sherry* (1969). "This book does not pretend to be an encyclopedia listing all European wines" (p. 13), and detailed information is still best found in Alexis Lichine's *Encyclopedia of Wines and Spirits* (see entry 23). Jeffs's book presents interesting historical and current information on pressing and fermentation, followed by details of soils, climate, vineyards, and wines in famous regions of France, Germany, Italy, Spain, and Portugal. The text is well footnoted, but there is no bibliography. There is an index of wines and regions. Sixteen maps indicating wine-growing areas are included. Jeffs's work presents histories of the first growths and offers much interesting information for those seeking more than a superficial survey of major wine-producing regions of Europe. This is another Faber book on wine.

138. Pogrmilovic, Boris, ed. **Wines and Wine-Growing Districts of Yugoslavia.** Zagreb, Zadruzna Stampa, 1969. 98p. illus.(part col.). maps (part col.). LC 73-974254.

This slim book is mostly filled with illustrations, but each district is carefully examined in view of the potential for tourists and the quality of the wine. Good sections on Merlot and Lujute wines.

FRANCE

Most of the words written about wines are about the products from France—and this often includes up to 85 percent of a general wine book that is supposed to cover other areas as well. Personal accounts also dwell on French wines. But it must be recognized that the best of the French wines is also the best of the world's offerings. Because it is a well-defined and important industry, the wine trade and viticulture in France has been the subject of many specialized studies, which are unknown to other countries. For instance, there is a mammoth three-volume work by Gaston Galtier (*Le Vignoble du Languedoc mediterranéen et du Rousillon*, Montpellier, 1960), crammed with illustrations, maps, colored charts, plans, tables, and diagrams. Or consider the Fédération Historique du Sud-Ouest and their *Vignobles et vins d'Aquitaine* (1970), a 500-page book that details the art and economic history of that area (fully illustrated). Joseph Capas presented *L'Evolution de la législation sur les appellations d'origine et la genèse des appellations contrôllées* (Paris, 1947) as the definitive history of government regulations of vineyards and wine production.

General

In the section below, wine books on specific areas in France are described, preceded by general works on all of France alone. For some unaccountable reason, there are no books in English on the Rhone Valley (covering Côte Rotie, Chateauneuf du Pape, Tavel, and Hermitage wines).

139. Churchill, Creighton. **A Notebook for the Wines of France.** New York, Knopf, 1961. 387p. illus. $8.50. LC 61-14192.

This is a wine diary or cellar book listing the 900 most important (in Churchill's opinion) French wines (and/or their vineyards) with space for the wine drinker's own records and notations. It covers all the principal wine districts of France, with tables explaining classifications in each of the respective districts. Samuel Chamberlain contributes an introduction.

140. Dion, Roger. **Histoire de la vigne et du vin en France des origines au 19e siècle**. Paris, 1959. 768p. illus. maps. price not reported. LC 60-22070.

This lengthy French epic is a comprehensive treatise on French vines and wines. Because of its full details on the vineyards and the major wine-producing areas, plus its historical matter, tables, maps, and excellent illustrations, this book is the most authoritative reference volume on the subject of French wines through the nineteenth century.

141. Dumay, Raymond. **Guide des vins de pays**. Paris, Stock, 1969. 313p. illus. 30 Fr.Fr. LC 73-476273.

This is the first real major French consumer guide to ordinary French wines, and that it appeared at all created some controversy. Even though it is written in French, it is designed to help both the native and the traveller explore the lesser known French wines. Wines are tasted and rated in a unique system.

142. Garvin, Fernande. **French Wine Handbook**. New York, Food from France, 1971. 62p. illus. maps. free.

A very broad and general book that is a good introduction to the field of French wines. It covers selection, parties, tastings, wine cellars, experts, advice and a short glossary. Each region is described, and comparisons are made between types of French wines so that one may be found for every occasion.

143. Hyams, Edward S. **The Wine Country of France**. Philadelphia, Lippincott, 1960. 208p. illus. $4.50. LC 60-12215.

This is an English novelist's travelogue through the vineyards of France; he speaks with great knowledge, since he himself is a wine grower with a vineyard in East Kent, England. Concentrating on the neglected areas of the Jura, Midi, and Corsica, he presents the economics of vineyards in their day-to-day operations. Other viticultural matters, such as pruning, are discussed. There are also many interesting asides on food and history.

144. Jacquelin, Louis, and René Poulain. **Vignes et vins de France**. Paris, Flammarion, 1970. 483p. illus. 68 Fr.Fr. LC 72-581555. English Ed.: **The Wines and Vineyards of France**. New York, Putnam, 1962. 416p. illus. $9.95. LC 62-14738.

The first book is the most recent and revised, while the second book is the only English version at present available. This is a collection of the names of all the vineyards of France (as well as of all the allowable wine grapes). Both authors are highly skilled oenologists. Introductory material includes a brief history of grape origins, followed by an outline of modern methods of wine growing and wine making (with an emphasis on soil and climate). Histories of *all* the major vineyards are given, plus the laws governing them. Geographic coverage also includes Corsica and Algeria. Over 5,000 vineyards are listed, and for each is given: the nature of the soil, the grape varieties planted, the yield of the wine, the official classifications of quality for each district, and up-to-date vintage guides. Miscellaneous information includes wine labels, measures, storage and service, wine and foods, and a wine vocabulary. There are 76 black and white photographs and 17 maps. The latest (French) edition is recommended for the wine student.

145. Lichine, Alexis, and William E. Massee. **Wines of France**. 5th ed. rev. New York, Knopf, 1969. 337p. illus. maps. $8.95. LC 71-79345.

This is a classic book that takes the reader on a tour of chateaux and vineyards, with re-evaluations of the vintages from 1929 to 1968. Descriptive material also includes information on markets, the personalities of the wine trade, and on the vintners themselves. Unfortunately, about 100 pages of this book (those concerned with the chateaux and vintages) duplicates Lichine's *Encyclopedia* (see entry 23). Also, at the same time as this edition appeared, the seventh edition was published in England by Cassell.

146. Ray, Georges. **The French Wines**. New York, Walker, 1965. 153p. illus. maps. $3.95. LC 65-15127.

First published by Presses Universitaires de France (1946), this American edition has been translated and updated by Paul Capon. This is part of a handbook series for the tourist, written in simple language. It covers the history of wines, region by region, and it also discusses eaux-de-vie and allied industries (such as corks, casks, and barrels). A snappy 18-page summary of French wines and appellations concludes the book.

147. Rowe, Vivian. **French Wines Ordinary and Extraordinary**. London, Harrap, 1973. 111p. illus. maps. $3.70. LC 73-155695.

Over 300 French wines, the majority of which are the lesser-known and cheaper brands, are itemized and briefly described. All are listed in the index and each is graded according to quality. This is a useful little book for the traveller in France and for those who wish to "shop around" at home. Route maps are given.

148. Shand, P. Morton. **A Book of French Wines**. 2nd rev. ed. Edited by Cyril Ray. Baltimore, Penguin, 1964. 278p. $1.50pa.

Originally published in 1928, this book was fully revised by Knopf (1960) before he died. It includes new chapters on the changed tastes in wine, the reorganization of French wine growing, and the overhaul of the legal machinery. It was further revised by Cyril Ray for the 1964 Penguin edition. This classic book is very complete; it goes rather deeply into country wines and brandies. The appendix contains a worthwhile summary of the "appellation contrôllée" laws. Unfortunately, there are neither illustrations nor maps.

149. Simon, André L. **The Noble Grapes and the Great Wines of France**. New York, McGraw-Hill, 1957. 180p. illus.(col.). maps (col.). $15.00. LC 57-9436.

Another in the posh McGraw-Hill series on wine books. Coming from Simon, this is instantly a definitive book. Many statistics are given, plus listings (by area) of wines in good vintage years. The obligatory color plates are immaculate, the maps good and the glossaries correct.

150. Vandyke Price, Pamela. **Eating and Drinking in France Today**. London, Tom Stacey, 1972. 324p. illus. maps. bibliog. £2.80. LC 72-190106.

This is a very welcome revised edition of the author's out-of-print *France: A Food and Wine Guide* (Hastings House, 1966). It is mainly a travel book for the tourist, but it also evokes fond memories for the armchair traveller. Part one details how the French eat; how to cope with French hotels; how to interpret a menu; and a general guide to food, wines, and aperitifs, including mineral waters, beers, and hot drinks. Other information includes manners and etiquette for the visitor and the preparations involved in picnics. Many of the drinks and much of the food are simply not available domestically, and a trip to France is a necessity for the gourmet.

Part two is the regional breakdown, with detail about visiting the wine area (from Alsace to Savoie). Subsections include specialities and dishes; cheese; local wines; things to see and do (based on gourmet tastes: museums dealing with food, wine or lifestyles; wine cellars; *objets d'art* to do with food and drink); and a list of hotels and restaurants personally recommended by the author. Throughout, there are glossaries and translation dictionaries after each chapter. There are no recipes, but the bibliography includes cookbooks, travel books and guides, and wine books. General updating is to 1972. A great book.

151. Warner, Charles N. **The Winegrowers of France and the Government since 1875**. New York, Columbia University Press, 1960. 303p. $6.00. LC 60-7130.

A specialized study in business history, this book details the trials and tribulations in combatting phylloxera with government assistance, the demarcation of growth areas, the involvement in tariffs and trade by the government, and the inside stories concerning the "appellation contrôllée" laws.

152. Wildman, Frederick S. **A Wine Tour of France**. New York, Morrow, 1972. 335p. illus. maps. $8.95. LC 70-182453.

The author, a wine importer, has written a guide to be used for visiting France and the vineyards. He suggests a three-week excursion, and provides a detailed itinerary, with recommended routes, lodgings, restaurants, mileage, telephone numbers, and names of proprietors (all but the latter are from the *Guide Michelin* and the *Guide Kleber*). Four 21-day motor trips are given through Champagne and Alsace; Burgundy and Rhone; Armagnac, Bordeaux, and Cognac; and the Loire Valley, Normandy, and Calvados. Each day has a separate map. Other information provided is the summary of vintages from 1959 to 1971 by region (with charts and ratings), plus the usual miscellaneous material on selection, storage, serving, and drinking. Great value for the wine novice who wishes to travel through France, but only the few pages detailing the itinerary will be of interest to the experienced hand.

153. Woon, Basil D. **The Big Wines of France**. Vol. 1. London, Wine and Spirits Publications, 1972. 204p. illus. £3.65.

This is a collection of "reports" from 11 wine areas of France, together with a glossary of wine terms ("Language of wine") and a pronunciation guide to wines and vineyards. The installments, now expanded, originally appeared in *Wine* magazine from 1970 to 1972. A second volume is projected.

Alsace

154. Hallgarten, S. F. **Alsace and Its Wine Gardens**. Rev. ed. London, Andre Deutsch, 1965. 196p. illus. £2.05.

This is an indispensable guide to the many varietal wines of Alsace, including a detailed journey along the "route de vin" (see also entry 156). Mr. Hallgarten is a scholar of German-type wines.

155. Layton, Thomas A. **Wines and People of Alsace**. London, Cassell, 1970. 209p. map. £2.50. LC 71-533869.

Building on Hallgarten's book, Layton describes the physical character of the people who inhabit the Alsatian Marchland, which has been either French or German for over a thousand years. All the wines are discussed, and the author points out that Alsatians can be divided into lovers of Riesling or lovers of Gewürtztraminer. Mr. Layton has also prepared similar books on the Loire and on Spain. (See entries 181 and 211.)

156. **La Route de vin: Alsace**. Paris, Editions d'Art les Heures Claires, 1966. 205p. illus.(col.). 150 Fr.Fr. LC 79-546552.

A very posh, colored-plate coffeetable book, with French texts by various authors (such as Paul Stehlin). The "route" from one vineyard to another is stunningly recreated by the gorgeous photographs. Certainly this is one for the armchair traveller.

Bordeaux

157. Allen, H. Warner. **Claret**. London, T. Fisher Unwin, 1924. 44p. maps. out of print.

This slim book includes a large fold-out map of the area and describes how claret (Bordeaux) is made. Little has changed in the past fifty years.

158. Carter, Youngman. **Drinking Bordeaux**. New York, Hastings House, 1966. 95p. illus. maps. (Drinking for Pleasure Series). $3.95. LC 66-7096.

As one of a series (see also entry 173), this book has more photographs than text. Here is a brief history of Bordeaux, including the heavy trade with England in the Middle Ages and the British possession of Bordeaux. While much material may be available elsewhere, there are over a hundred photographs, many by the author, that make this book a bargain.

159. Cocks, Charles, and Edward Feret. **Bordeaux et ses vins: classé par ordre de mérite**. 11th ed. Bordeaux, Feret et Fils, 1949. 1132p. illus. maps (col.). 100 Fr.Fr.

This is a standard and invaluable reference work on the chateaux of Bordeaux. While written in French the material is easily retrieved because of its tabular nature. Each chateau is fully described. First edition was in English in 1846; the last English edition was the third, in 1899. Also available in German.

160. Penning-Rowsell, Edmund. **The International Wine and Food Society's Guide to the Wines of Bordeaux**. New York, Stein and Day, 1970. 320p. maps. bibliog. $10.00. LC 77-106797.

Part one is an historical survey of the development of Bordeaux, with a history of the wine merchants and a history of the Médoc classifications. The author examines the importance of Bordeaux and its trade to England. Part two describes the vines growing in each district and the vinification process. Supplementary material details the opening wine prices for Bordeaux since 1831, the vintages since 1890, and the annual rainfall.

161. Ray, Cyril. **Fide et Fortitudine: The Story of a Vineyard**. London, Pergamon Press, 1972. illus. £2.50.

This corporate history was written to commemorate the 150th anniversary of Barton's acquisition of Ch. Langoa and Leoville. The slender but lavish work gives much historical background and the story of the Barton family, including some material on Burgundy and Rhone acquisitions through natural expansion.

162. Ray, Cyril. **Lafite: The Story of Château Lafite Rothschild**. New York, Stein and Day, 1969. 162p. illus. facsim. bibliog. $6.95. LC 69-17938.

Ray, the author of other company histories, has produced this memoir of Lafite's first hundred years. At the time, it was the first English book ever to be devoted to a single vineyard and its wine. This is the most famous wine in all the world, and this fame contributes to the Second Empire nature of the Lafite vineyard. A well-written account brimming with biographical detail and social history.

163. Roger, J. R. **The Wines of Bordeaux**. New York, Dutton, 1960. 166p. illus. maps. bibliog. $3.75. LC 60-52297.

This comprehensive book presents an opening chapter detailing the general characteristics of Bordeaux, along with summary data that are almost tabular in construction. Succeeding parts deal with the 1855 classification, and 13 chapters cover the general appellations of the area. A section called "wine and food" tells what wine supposedly goes with what food. The short bibliography deals only with French books.

164. **La Route du vin de Bordeaux**. Paris, Editions d'Art les Heures Claires, 1968. 173p. illus.(col.). 150 Fr.Fr. LC 75-548477.

A very posh, colored-plate coffeetable book, with French texts by various authors, such as Georges Portman. The "route" from one vineyard to another is stunningly recreated by the gorgeous photographs. Certainly this is one for the armchair traveller.

Burgundy

165. Brejoux, Pierre. **Les Vins de Bourgogne**. Paris, Atlas de la France Vinicole L. Larmat; Société Française d'Editions Vinicoles, 1967. 276p. illus. maps. 24 Fr.Fr. LC 68-133833.

A superb reference work by the author of *Les Vins de Loire* (see entry 180). Well written in a colloquial French style, individual chapters carefully detail the area with respect to geography, production, appellation contrôllée laws, histories, and statistics of production. This is an extremely detailed book about vineyards, classifications, and productions done for *La Revue du vin de France*. The nine maps are excellent.

166. Gadille, Rolande. **Le Vignoble de la Côte Bourguignonne**. Paris, Les Belles Lettres, 1967. 688p. maps. tables. bibliog. (Dijon University Publications, 39). 50 Fr.Fr. LC 68-117516.

This learned dissertation is a model of its kind, and it is needed for *all* areas of grape and wine production the world over. It deals with climates, micro-climates (temperatures, isolation, relative humidity, relation of Dijon area to the rest of Burgundy), morphological and topological structures, soils (plateau, plains, sloping), erosion, rainfall, sunshine, plant life and vegetation, and the evolution and history of vine and wine. The author attempts to determine what factors create the best wine. The wide fluctuations of soil and climate are examined, and comparative investigations are made into the years of best quality (1949 and 1959), good quality (1948 and 1950), and mediocre quality (1951 and 1954). Also covered are the human elements and the influences of modern techniques. By examining the physical and human foundations of a high quality viticulture, the author tries to answer the question "what makes Burgundy wine so fantastic and perfect?"

167. Grivot, Françoise. **Le Commerce des vins de Bourgogne**. Paris, Sabri, 1964. 224p. illus. maps. bibliog. price not reported. LC 65-76116.

The author describes (in French) the impact of Burgundy wine on the economy of France, and presents a detailed outline of the A.O.C. laws of 1905, 1919, and 1935. The work examines the structure of the wine industry, and the roles played by the grower, the broker, the shipper and merchant, and the relevant associations. The importance of the banker and the storage warehouses should not be underestimated. There is also a discussion of how wine prices are fixed by the Hospices de Beaune auctions. The pricing for tonneaux is broken down by region. There is a good 160-item bibliography, but unfortunately the book lacks an index.

168. Poupon, Pierre, and Pierre Forgeot. **Les Vins de Bourgogne**. 6th ed. revue et mise à jour. Paris, Presses Universitaires de France, 1972. 221p. illus. maps. 13 Fr.Fr.pa. LC 72-334023.

Although this work is revised every four years or so, the current English edition is unfortunately only the second (1959) edition (New York, Hastings House). This is a very authoritative book on the burgundian wines, their appellations, and the comparative rankings from Chablis to the Beaujolais.

169. Rodier, Camille. **Les Clos de Vougeot**. Dijon, L. Damidot, 1959. 173p. illus. maps. price not reported.

This reprint of a 1931 classic is a solid history of one of the Côte d'Or's oldest and most celebrated vineyards. Many anecdotes and much wine lore abound in the text, which is liberally sprinkled with prints, maps and drawings.

170. Rodier, Camille. **Le Vin de Bourgogne**. 3rd ed. Dijon, L. Damidot, 1948. 351p. illus. maps (part col.). price not reported.

This French text is a standard reference work for the wines of the Côte d'Or. Each vineyard of significance is given a short history, and there are many tables of statistics dealing with production and consumption.

171. **La Route du vin de Bourgogne.** Paris, Editions d'Art les Heures Claires, 1970. 175p. illus.(col.). 150 Fr.Fr. LC 75-547260.

Another very posh, colored-plate coffeetable book, with French texts by various authors, such as Jacques de Lacretelle. The "route" from one vineyard to another is stunningly recreated by gorgeous photographs. This is certainly one for the armchair traveller.

172. Yoxall, Harry Waldo. **The International Wine and Food Society's Guide to the Wine of Burgundy**. New York, Stein and Day, 1970. 191p. illus. maps. bibliog. $7.95. LC 70-106798.

In addition to the reds, whites, rosés and sparkling burgundy, marc is also covered. Mr. Yoxall visits the domaines, the auctions, and the tastings, and he eats the classic meals. His liberal (at least, legal) definition of Burgundy also includes Chablis and Beaujolais. In this context he deals with "fake" Burgundy wines. Also, there is a discussion about prolonged vinification versus short vinification. Seven appendices cover a vintage chart, classifications, listings (by "grand" and "premier" crus) plus material on the A.O.C. minimum and maximum quantities in the vinification process.

Champagne

173. Carter, Youngman. **Drinking Champagne and Brandy**. New York, Hastings House, 1967. 96p. illus. maps. (The Drinking for Pleasure Series). $3.95. LC 68-618.

Here is basic history of champagne before Dom Perignon, and a past history of cognac. Considering the length of the book, too much was attempted on Armagnac and brandies (e.g., calvados, quetsch, and marc). Large type and many pictures further reduce the amount of text, and much material that *is* provided is available elsewhere. Yet over 100 photographs, many by the author, count for a lot, and it is the pictures that make this book a bargain.

174. Comité Interprofessionnel du Vin de Champagne, Epernay. **Champagne: Wine of France**. 2nd ed. Paris, L'Allemand Editeur, 1968. 48p. illus.(chiefly photos). free.

This book covers much the same ground as Carter's (entry 173). The English edition presents the basic history, with many photographs of tapestries and vessels. Included are descriptions of wines, vineyards, cellars, cultivation, harvesting, pressing, the cuvée, fermentation, bottling, corking, and so forth.

175. Forbes, Patrick. **Champagne: The Wine, the Land, and the People**. New York, Reynal, 1968. 492p. illus. maps. tables. bibliog. $10.00. LC 67-106422.

Forbes presents a unity of subject matter: Champagne refers to the province, the wine-field within, and the wine itself. Hence, in this epic, which is the result of nine years of work, Mr. Forbes deals with wine, countryside, soil, geological features, history, and people. He opens his book with a travelogue, then a history (Champagne is the cross-roads of Europe), followed by a discussion of sparkling champagne and its modern history. Part two is more technical, dealing with how the wine develops (how it is planted and tended, the annual cycle of work), and a description of the harvesting and vinification. There are also the obligatory tips on drinking, anecdotes, points about people and food, material on bottles and corks (very important), and chapters on the 15 great champagne makers (corporate history type). The six-page bibliography, when checked against that maintained by the grower association, is fairly complete.

176. Gandilhon, René. **Naissance du Champagne, Dom Pierre Pérignon**. Paris, Hachette, 1968. 288p. illus. 100 Fr.Fr. LC 70-376792.

As another lavishly illustrated book on champagne, this work concentrates on the discovery of the champagne secondary fermentation process, the exciting role played by Dom Pérignon, and the use of corks in champagne bottles.

177. Ray, Cyril. **Bollinger: The Story of a Champagne**. London: P. Davies, 1971. 179p. illus. maps. £3.25. LC 70-859830.

This is a corporate history of a champagne manufacturer, by an author who is skilled in such writings (see entries 161, 162). Apart from the Bollinger firm itself, the highlight here is the fine detail that Ray presents (e.g., his description of the story of the 1811 Champagne riots).

178. **La Route du vin de Champagne**. Paris, Editions d'Art les Heures Claires, 1966. 189p. illus.(col.). 150 Fr.Fr. LC 70-551600.

A very posh, colored-plate coffeetable book, with French texts by various authors such as Armand Lanoux. The "route" from one vineyard to another is stunningly recreated by the gorgeous photographs. Also valuable is the display of 20 lithographs of Touchagnes.

179. Simon, André L. **The History of Champagne**. New York, Octopus, 1971. 195p. illus. map. bibliog. $5.95.

Originally published in 1962 as one of the McGraw-Hill wine series, this book has been inexpensively reprinted. This is a solid and basic history of champagne and of the various theories of how the wine came to be. Color plates provide good illustrations, and the text is interlarded with many statistics, including a table of vintage years from 1800 to 1861. There is also a chapter on "American Champagne" (totally uncalled-for) by Robert Misch.

Loire

180. Brejoux, Pierre. **Les Vins de Loire**. Paris, Compagnie Parisienne d'Editions Techniques et Commerciales, 1956. 239p. illus. bibliog. (Cuisine et Vins de France). 100 Fr.Fr. LC A59-283.

The French text describes the countryside, the wine grower, and the wines, with good sections on the *Sauvignon blanc* grape (Pouilly sur Loire, Sancerre), the Touraine area (Vouvray, Montlouis, and the red wines of Chinon and Bourgeuil), and other wines such as Anjou, Saumur, and Muscadet. This super guidebook delves into food and presents information on a cross-section of growers.

181. Layton, Thomas A. **Wines and Châteaux of the Loire**. London, Cassell, 1967. 225p. illus. map. £5.25. LC 67-89262.

Although this is similar to Layton's other books on the countryside and wines of Alsace and Spain, he is here given more room by the very nature of the chateaux along the Loire. Sixteen plates present the best chateaux (also found elsewhere in more significant art and architecture books), but the wine sections are important because this book is in English.

182. Raimbault, Henri. **Les Vins d'Anjou et de Saumur**. Angers, Impr. S.E.T.I.G., 1967. 104p. illus. 11 Fr.Fr.pa. LC 68-103680.

This promotional book for the tourist in the Loire areas is merely a catalog of wines available in the area.

GERMANY

German wines are well known in America, but only recently have they begun to sell well; that market may soon collapse, however, because of the re-evaluation of the mark. Germany is a major wine-producing country with great wines, and it is rather surprising that there are so few books (and of such pedestrian quality). What follows is a selective list of the best, plus a German ringer that is a classic production. Other German books of note that are sterling examples of what *could* be done are G. Stein's *Reise durch die deutschen Weingarten* (Munich, 1956) and G. Leonhardt's *Weinfachbuch*

(Leipzig, 1954), both still extensively used in the German wine trade. Another problem of discussing German wines is that in 1971 the wine laws were so radically changed that the cumbersome and mysterious (some would say notorious) German wine label, with all its puzzling nomenclature, disappeared for all wines produced after the 1971 harvest. By the end of 1974, most of the older wine stocks will have disappeared, so the labels will be easier to read; only three components need be identified: type of wine, place, and quality. Only one recent wine book has caught these changes, and a handful of general surveys have incorporated them into the text as an addendum.

183. Andres, Stefan. **Die grosse Weine deutschlands**. Berlin, Verlag Ullstein, 1960. 190p. illus. bibliog. LC A61-3540.

This beautifully illustrated book, with German text by a well-known German authority, details the best wines of that country. The author carefully explains the process of Auslese, Spatlese, Eiswein, and the production of Trockenbeerenauslese wines. Many recommended (and excellent) vineyards are listed here which do not appear in the majority of the English-language books on German wines.

184. Hallgarten, S. F. **Rhineland, Vineland: A Journey through the Wine Districts of Western Germany**. 4th ed. rev. and enl. London, Arlington, 1965. 319p. illus. maps. bibliog.

This is a classic book that was originally published in 1951. It contains superb black and white photos and decent maps, plus many personal accounts. The extensive bibliography makes reference to many German works. The first third of the contents covers the history of German viticulture, the grape harvest, sweetening, fermentation, filtering, blending, the pre-1971 German nomenclature, diseases of the vines, judging and tastings, and when to drink German wines. The next third concentrates on touring each of the ten major areas–Rhinehessia, Wahe, Rheingau, Mittelhein, Moselle, Franconia, etc.–with descriptions of German "Sekt," wine festivals, and local cuisines. The last third of the book has 13 appendices dealing with German wine-growing centers, vineyards, class names and districts, tasting notes, glossaries, and wine classifications.

185. Langenbach, Alfred. **German Wines and Vines**. London, Vista Books, 1962. 190p. illus. maps. bibliog. £1.75. LC 64-6702.

This short survey, by a leading member of the German wine trade, is designed primarily for students professionally concerned with the subject, and secondarily for amateurs. He opens with a brief history of German viticulture, with two technical chapters on the different types of vines grown, the methods of cultivation and the gathering or harvest. Other chapters describe the production treatment until the wine is bottled. The subject of difficult wine names is covered, but, of course, the 1971 wine laws have changed all that. Fifty-five pages deal with an extended survey of the 10 main districts, with lists of the leading sites. There is also a chapter on German "Sekt," the sparkling wine.

186. Loeb, O. W., and Terence Prittie. **Moselle**. London, Faber and Faber, 1972. 221p. illus. maps. bibliog. £4.50. LC 72-169958.

One of the Faber series of wine books, this is the only modern wine book in English that deals with one particular section of German wines. It is a complete book, with maps and photographs, that presents concise information on all the vineyards in the Moselle Valley. Covered are the principal vineyards of the Upper Moselle, Saar, Ruhr, and Middle and Lower Moselle Valleys. Both authors draw on personal experience and the historical record. Loeb is a wine merchant born in the Moselle area, and Prittie is an experienced wine journalist.

187. Meinhard, Heinrich. **German Wines**. Newcastle-upon-Tyne, Oriel Press, 1971. 102p. illus. maps. £1.75. LC 70-135981.

"Branded wines with fancy names" are austerely excluded from this anthropologist's study of German wines. He believes good wine is rare and worth the search. He gives a concise description of each region, an account of 17 district grape varieties, and a short history that goes back to the second century (A.D.) "Romanized" Celtic wine growers. Additional material includes short chapters on the new 1971 wine law, modernization, and advice on selecting from a German wine list. An incredible amount of information sandwiched into 100 pages.

188. Rudd, Hugh R. **Hochs and Moselles**. London, Constable, 1935. map. 165p. out of print.

This is an unillustrated but literate tour of the areas of the Palatinate, Moselle Valley, Saar, Nahe, Rheingau, and Rheinhessen. More a series of personal accounts and comments on the vintages, and hence suited for advanced study. This volume of the Constable Wine Library is well worth seeking out.

189. Schoonmaker, Frank. **The Wines of Germany**. Rev. ed. New York, Hastings House, 1966. 156p. illus. $2.95pa. LC 66-8909.

Schoonmaker probably knows more about German wines than any other English-speaking individual. This classic study, originally published in 1956 and updated only as to vintage years and changes of ownership in vineyards, comprises mainly reprints from his *Gourmet* magazine articles. He opens with a discussion on geography and wine types, and then describes the four major growing areas–Moselle, Rheingau, Rheinhessen, and Palatinate. Along the way he discusses growers and presents excellent sketch maps. Concluding material includes how to buy and store German wines, and how to serve and taste them. Quite properly there are only two pages on red wines, but there is only one page on Sekt, which is now quite good and which deserves more space. Much information is given on interpreting the wine label, but this is now obsolete for current practice. An unusual section at the end is "a brief list of German wine-tasting terms," not found in any of the other books in this section.

190. Simon, André L., and S. F. Hallgarten. **The Great Wines of Germany and Its Famed Vineyards**. New York, McGraw-Hill, 1963. 197p. illus. plates (col.). maps (col.). tables. $12.50. LC 63-19442.

About a decade ago, McGraw-Hill produced an exceptional series of wine books, all well illustrated and all overpriced—for the coffeetable. With color photographs and maps, this one was another winner, yet it duplicates much of the information in other books on German wines. More than a third of this volume contains production statistics.

GREAT BRITAIN

191. Launay, André. **The Merrydown Book of Country Wines**. London, New English Library, 1968. 118p. £0.30pa. LC 78-432777.

This is a brief history of all types of wines in England, and of the grape vine, with concentration on fruit wines: apple cider, mead, currant, elderberry, gooseberry, orange, herbs, and vegetables. There is also a short history of the Merrydown Company, and some recipes for punches and wine cups.

192. Rook, Alan. **The Diary of an English Vineyard**. London, Wine and Spirit Publications, 1972. 124p. illus. £0.90. LC 72-193828.

This is the story of the creation and development of a small vineyard in Lincolnshire, with a detailed account of the difficulties and rewards in producing the vines of the early years.

193. Simon, André L. **English Wines and Cordials**. London, Gramol, 1946. 144p. illus. LC 47-28241. out of print.

Much of this little book is about the fruit wines of England; it also covers the concoctions made of syrups and alcohol, punches, and mixed drinks.

ITALY

194. Dettori, Renato G. **Italian Wines and Liqueurs**. Rome, Federazione Italiana Produttori ed Esportatori di Vini, Liquors ed Affini [FEDER-VINI], 1953. 158p. illus. free.

This description of Italian wines is in English. It is a highly idealistic history of wine, but readable even if it is a little flowery. There are superb watercolor illustrations, with good notes on the home use of wine (storage, decanting, serving, etc.). One unique and very good feature is a sort of reverse wine glossary. Instead of "bright," "ruby," etc., there are categories for grading, with the descriptions within to verbally assess the wine. Thus, there are separate sections for: clarity, color, sweetness, acidity, alcoholic content, texture, bouquet and taste, and body. Liqueurs discussed include Italian brandy and herb liqueurs. The appendix contains many cocktail recipes, a high percentage of which use Italian brandy (which is suitable for anything but neat drinking); a complete list of Italian wines whose collective names of origin are protected by law; and a list of producers and exporters.

195. **Discovering Italian Wines: An Authoritative Compendium of Wines, Food, and Travel through the Nineteen Producing Regions of Italy**. Prep. under the auspices of the Italian National Wine Committee and the Italian Foreign Trade Institute. Los Angeles, Ward Ritchie Press, 1971. 136p. illus.(col.). $7.95. LC 70-133249.

After a foreword by Robert Baizer and five pages devoted to the "history and legends" of wine, the book consists of two main sections: "The Controlled Wines of Italy, by Region" and "The Foods of Italy." The controlled wines of each region, in the first of these sections, are described in about a paragraph, and their "properties and characteristics" are defined (color, "taste," and alcoholic content). The perils of trying to describe the taste of a wine are obvious (many wines, it seems, are "soft, harmonious, and dry"—or some inversion of these). In the second section, regional recipes are presented, with a comment on each by Mike Roy (former emcee for "Duffy's Tavern"). The recipes are pretty elementary (e.g., one ingredient is listed as "tomatoes, pear shaped," and the overall format is similar to Veronelli's earlier book (see entry 200). But the color photographs are very nice.

196. Layton, T. A. **Wines of Italy**. London, Harpers, 1961. 231p. illus. £1.50.

A complete book covering all the wines normally available in the British market. Yet it is excessively anecdotal, and it pales in comparison to Ray's book (entry 198).

197. Paronetto, Lamberto. **Chianti: The History of Florence and Its Wines.** Verona, Enostampa, 1969. 224p. illus. bibliog. £2.50.

Originally published in 1967, this is mainly a travel book that investigates Tuscany, its past and present. There are numerous color and monochrome illustrations.

198. Ray, Cyril. **The Wines of Italy**. Rev. ed. Baltimore, Penguin Books, 1971. 211p. illus. maps. $1.45pa. LC 72-181233.

This is an index of principal wines. Arrangement is by region (with preliminary material), followed by the types of wines found locally. This dictionary form results in over 600 entries. The introductory matter outlines the history of Italian wines. First published in 1966, it was immediately translated into Italian and won a Bologna Trophy in 1967. The 1966 McGraw-Hill hardcover version is still available at $12.50; in view of the substantial price difference, it is recommended only if the reader wishes the following, which are *omitted* from the paperback: lavish photographs; a long appendix detailing the 1965 wine laws of Italy; statistical tables; and notes on vintage years for 50 of the 600 wines listed. The paperback has only one addition for the 1971 "revised" printing: a 12-page amendment to the introduction. This is a *very* useful book on Italian wines, but unfortunately no more than a couple dozen of the wines are available in either Britain or the United States. Most Italian wines stay at home, and this book is good for the curious traveller or the resident abroad.

199. Roncarati, Bruno. **D.O.C.: The New Image for Italian Wines**. London, Harpers Trade Journal, 1971. 89p. maps. £0.75. LC 72-172597.

This is a detailed description of quality Italian wines, with particular reference to the new laws on controlled denomination of origin. He explains all the new regulations since 1963. This is an exceptionally well-detailed, well-written account covering the main geographic areas. The author translated Paronetto's work (see entry 197).

200. Veronelli, Luigi. **The Wines of Italy**. New York, McGraw-Hill, 1965. 326p. illus.(col.). LC 64-19614. out of print.

It is rather surprising that McGraw-Hill would publish *two* books on Italian wines within one year of each other, both about the same size, with lavish photographs; to compound the confusion, both are "reference" type books with a high density of data per page. While Ray's book investigated 600 types of wine, in dictionary form, Veronelli presents his similar information in tabular or directory form. After 80 pages of general information (selection, storage, serving), he presents a directory of major producers, by region and town, with sample labels (actual labels are tipped in). Material, arranged by the types of grapes used, describes the grape and the wine, the alcoholic content, the acidity, the use and the service. It is perhaps a bit more practical than Ray's book, although it should be noted that both mention many, many wines not available in North America. And perhaps because Ray is easier to read (Veronelli's is one to study), this present book has been allowed to go out of print.

PORTUGAL

201. Allen, Herbert Warner. **The Wines of Portugal**. New York, McGraw-Hill, 1964. 192p. illus.(col.). maps (col.). $10.00. LC 63-22124.

Another volume of the McGraw-Hill series, characterized by large type and colored plates (24 of them), this work devotes most of its space to port (and vintage port, at that), with, as Mr. Allen put it, the "history of secrets won and lost." Yet the range is wide. There is a discussion of aperitifs, dessert wines, red, white and rosé table wines, and, of course, madeira. Portugal is a wine-producing country noted for its wide range of offerings in alcoholic beverages. Mr. Allen's style is comfortable—far more so than others in the series—and he includes vintage charts for ports.

202. Bradford, Sarah. **The Englishman's Wine: The Story of Port**. New York, St. Martin's Press, 1969. 208p. illus. bibliog. LC 71-446310. out of print.

A solid historical work (it goes back to the Douro in 137 B.C. for the founding of Oporto), this book is really the history of an area and its trade rather than the story of wine. There is a detailed description of the port trade today, including how port is made, important information about individual shippers and their practices, plus how "vintage" port is determined and the difference between Oporto-bottled and foreign-bottled port. Appendices describe the vintage years since 1900, and the British Port Houses (importers and merchants); these supplement Charles Sellers' *Oporto Old and New* (London, 1899), which described the genealogy of the "portocracy." The bibliography and the glossary of port terms cover two pages each. The somewhat small, dense text includes extensive quotes from historical correspondence.

203. Postgate, R. W. **Portuguese Wine**. London, J. M. Dent, 1969; distr. by International Publications Service, 1970. 102p. illus. maps. $6.25.

The main emphasis in this book is on table wines, as port and madeira are already covered so well (although there are short chapters on these two drinks). There is advice on what wines to drink, and where to drink them, with separate material on the green wines of the Minho district, the Douro clarets and white wines, the prolific but quality Dao wines, the wine of

Lisbon (great in Victorian days), and the Lagoa wines of the far south. Each area is illustrated by a map showing the numbered roads, and accompanied by notes on suggested routes. This book, then, can serve as a tourist guide. Each area is described as to topography, local history, origins, and growth and quality of wines. An individual chapter discusses tours of the wine-growing area generally. Concluding information covers the selection of Portuguese wine and the question of who imports what into Britain. All wines noted in the book were tasted by the author.

204. Sanderson, George G. **Port and Sherry**. London, Sanderson and Sons, Co., 1955. 98p. price not reported.

This book, mostly promotional, deals with those brands and labels imported into Britain by Sanderson and Sons.

205. Stanislawski, Dan. **Landscapes of Bacchus: The Wine in Portugal**. Austin, University of Texas Press, 1970. 210p. illus. maps. bibliog. $7.50. LC 79-96691.

A very readable account of the spread of viticulture in Portugal, mostly historical, with some technical aspects. The author examines the prime growing areas and tries to account for the role played by wine in the agricultural economy. Much statistical information is used.

206. Valente-Perfito, J. C. **Let's Talk about Port**. Oporto, Instituto do Vinho do Porto, 1948. 100p. plates. maps. price not reported. LC 49-22374.

This is a basic, promotional history and description of the Douro Valley, where port is vinified, and of the great wine houses in Oporto where it is aged and blended.

SPAIN

207. Croft-Cooke, Rupert. **Sherry**. New York, Knopf, 1958. 232p. out of print.

This historical account of sherry quotes at length from Gordon's book, originally published in Spanish in 1948 and now translated into English (see next entry). On its own, Croft-Cooke presents a good account of the solera system.

208. Gordon, Manuel Gonzalez. **Sherry: The Noble Wine**. New York, A. S. Barnes, 1970. 240p. illus. maps. $7.95.

This classic study was originally published in Spanish in 1948, and is here translated by Floyd M. Dixon. Gordon was active for 50 years in the family firm of Gonzalez Byass. Everything that an enthusiastic wine amateur would want to know about sherry is contained between these covers: vineyards, grapes, flor, soleras, two centuries of sherry trade, plus detailed information about the town of Jerez itself. Extensively illustrated with 32 pages of illustrations, 18 line drawings and two maps.

209. Huetz de Lemps, Alain. **Vignobles et vins du nord-ouest de l'Espagne**. Bordeaux, Les Impressions Bellenef, 1967. 2 vols. illus. maps. bibliog. (Madrid. Ecole des hautes études hispaniques. Bibliothèque 38). price not reported. LC 67-111369.

This work is offered as a typical example of the scholastic studies of viticulture being undertaken in France. This descriptive dissertation of a thousand pages, in French for the moment, is based on 12 years of research over 300,000 hectares (one-fifth of the Spanish wine production). Yet, as the author points out, the northwest area is large, and more quality wines, such as Rioja and Navarre, can result to quench the thirst of an expanding wine appreciation audience. This is an exhaustive study of a micro-climate, worth seeking out for the quantity of fresh data, and as a model for work to be done on all wine-producing areas of the world. It deals with the types, geography, and the basis for growing wine grapes. The history of vineyards since Roman times, up through the phylloxera crisis, and recent changes are covered. Detail is amassed on the working and production of vineyards; the owners and producers (the business end), with special emphasis on the small owner; the production of special wines and eaux-de-vie; and the development of Bodegas and cooperatives. There is an extensive 40-page bibliography.

210. Jeffs, Julian. **Sherry**. 2nd ed. London, Faber and Faber; distr. by British Book Centre, 1970. 283p. illus. map. bibliog. $9.50. LC 77-576368.

That there is much material available in Spanish about sherry is demonstrated by Mr. Jeffs' lengthy bibliography. The first part here is historical, based on original sources in Spanish and English and emphasizing the wine trade and politics of the two countries. Some descriptions are given of important wine merchants of the time. Part two is descriptive of techniques, the importance of flor and the solera system. Other scattered material deals with Manzanilla, blending, choosing a sherry (from fino to cream), statistical data on exports, the Guy-Lussac scales, and other bits of information that have never been published before. Illustrations comprise only eight plates (one in color), six diagrams, and one pithy map. Yet the O.I.V. gave this book the biennial prize for historical works in 1962 (the highest award for a book on wine). Jeffs, a former editor of *Wine and Food*, rewrote this book (and, of course, updated it) for its appearance as the first in the Faber series of books on wine.

211. Layton, Thomas Arthur. **Wines and Castles of Spain**. New York, Taplinger Publishing Co., 1960. 246p. $4.95. LC 60-9129.

Another part-travel, part-information book by Layton, who has also produced similar works on Alsace and the Loire Valley. Most of the material is historical.

212. Rainbird, George M. **Sherry and the Wines of Spain**. New York, McGraw-Hill, 1966. 224p. illus.(col.). maps (col.). $13.95. LC 66-25521.

Another "definitive" book in the McGraw-Hill series, lavishly illustrated, and complete with vintage charts, historical commentaries, and visits to the vineyards.

NORTH AMERICA

UNITED STATES

Of all the American wines, those of California get a fair slice of the wordage in most wine books. The rest of the hemisphere does not fare too well—especially South America, for which the definitive book (indeed, *any* book) has yet to be written. The main thrust of wine books, according to Adams, below, is to sell wine, and this must be so because South American wines have not penetrated the British market at all and are just becoming known in the United States, where they have been imported cheaply. Thus, the first book may well appear next year. As with most sections of this book, only the best or current items dealing with North American wines are listed here.

213. Adams, Leon D. **The Wines of America**. San Francisco, San Francisco Book Co.; distr. by Houghton Mifflin, 1973. 465p. maps. $10.95. LC 77-75347.

Adams, founder of the California Wine Institute, has spent half a century travelling and tasting wines. Upset by wine merchants who write books favoring their European wine imports over the domestic variety, he has produced this historical and descriptive work after seven years of hard labor. Starting from Scuppernong country (southeastern United States), where wine has been made for 400 years, he travels through the mid-Atlantic states, Ohio, New York, the Finger Lakes, New England, Michigan, California (greatest detail here), northwestern United States, Canada, and Mexico. Other material includes the rise of amateur winemakers and kosher winemakers. There is much detail on the wineries visited (and whether or not they are open to the public), as well as tastings at the winery itself. Concluding chapters describe wine taste, the impact of French hybrids, and a chronology of wine history in North America. The detailed index permits quick retrieval of comparisons, and the book is up to date (to the end of 1972). There are 11 inconsequential distribution maps, and there are no illustrations and no bibliography.

214. Balzer, R. L. **California's Best Wines**. Los Angeles, Anderson and Ritchie, 1948. 153p. $4.00. LC 48-9511.

An older book that takes the reader on a tour of the vineyards, a dinner at the Hearst castle in San Simeon, various wine tastings, the production of sherry, and recipes—all in Balzer's jaunty, popular style.

215. California. Wine Advisory Board. **California's Wine Wonderland: A Guide to Touring California's Historic Grape and Wine Districts**. San Francisco, 1972. 30p. maps. free.

The introduction discusses the characteristics of vintners and wineries, tastings, cellarage and restaurants and gives a mileage chart for motorists. Arrangement is by the nine major wine-producing areas, with the name, address, phone number, hours of opening, and public facilities (picnic facilities, retail sales, tastings) indicated. Almost all of the California vintners are here, as the Wine Institute represents over 90 percent of the yearly crush.

216. **California Wine Country**. By the Sunset Editorial Staff. Book Editor: Bob Thompson. Menlo Park, Calif., Lane Books, 1968. 96p. illus. maps. (Sunset Travel Books). $1.95pa.; $3.25 library binding. LC 68-26324.

This large-format book travels through California by region, from north to south. The maps have accompanying tourist information for the wineries located within the area shown. Information is given about tours, picnic areas, and so forth. Yet much here is out of date (by at least seven years), and current information is available (free) from tourist sources. What Sunset has done is to collate the information. But the book is cheap, the pictures are good, and there are suggestions for picnic food, bicycle tours, and tasting notes.

217. Crosby, Everett. **The Vintage Years: The Story of High Tor Vineyards**. New York, Harper and Row, 1973. illus. $8.50. LC 72-9113.

For 22 years, Crosby and his wife ran a winery about 35 miles up the Hudson River. This is the story of their adventures in keeping the operation going. Interesting information includes the vagaries of wine, weather, hired help, and the federal and state liquor commissions. The meat of the book documents the selection of hybrid varieties, their proper cultivation and harvesting, and the processes involved from grape to bottled wine.

218. Farrell, Kenneth R. **World Trade and the Impacts of Tariff Adjustments upon the United States Wine Industry**. Berkeley, California Agricultural Experiment Station, 1964. 114p. illus. bibliog. (Giannini Foundation Research Report, 271). LC 64-63730.

A cogently argued book that rests on the facts: three-fourths of the world's wine is produced in economically depressed Mediterranean areas, and there is much disposal of wine in the United States because of its low tariffs. Thus, to protect the domestic industry against fluctuating sales, the author is against tariff decreases. He says nothing about tariff increases.

219. Fisher, Mary Frances Kennedy. **The Story of Wine in California**. Foreword by Maynard Amerine. Berkeley, University of California Press, 1962. 125p. illus. $15.00. LC 62-18711.

Half-historical and half-technical, this book has superb photographs by Max Yavno. Yet in view of recent developments and historical findings, this book badly needs revising–and a lower price.

220. Greyton H. Taylor Wine Museum. **Living Wine Grapes**. Hammondsport, N.Y., 1972. 55p. illus. $1.00.

This little book, produced by the museum and Bully Hill Vineyards, describes and illustrates each variety of grapes grown in the Finger Lakes area, giving the grape's pedigree, history, average harvesting date, and the type of wine it produces. This is especially valuable for the grape grower and/or the amateur winemaker.

221. Massee, William E. **McCall's Guide to Wines of America**. New York, McCall, 1970. 210p. $6.95. LC 71-122147.

A wine guide for the novice, this book includes a short primer on kinds of wine, wines and winemakers of America by region, brief chapters on sparkling, fortified, and aromatic wines, an appendix with glossary, information on storing wines, and how to obtain hard-to-find wines. Two maps showing the location of wineries are provided, one for California and one for the Eastern states. The title is somewhat misleading, since almost half the book is devoted to California wines, making it similar to Melville's *Guide to California Wines* (next entry). The author jokes about the possibility of vineyards and winemaking in Canada and devotes two paragraphs to Argentina and a page to Chile, the two biggest wine producers in the hemisphere. Summing up, this useful guide for novice wine enthusiasts offers an uncomplicated introduction to basic information on California and eastern United States wines.

222. Melville, John. **Guide to California Wines**. 4th ed. rev. by Jefferson Morgan. San Carlos, Calif., Nourse Publishing Co., 1972. 233p. illus. $5.95; $2.95pa. LC 72-190042.

Since first published in 1960, at 235 pages, it has undergone revisions every four years (in presidential election years) and in the process it has lost two pages. Only minimal updating was needed for this fourth edition, so hang on to the third, if you have it. Part one of this reliable guide details the types of wines and vines available from California, and to some extent this is duplicated by Blumberg's book (entry 484), which deals rather extensively with tastings. Part two tells about the more notable wineries.

223. Norman, Winston. **More Fun with Wine**. New York, Pocket Books, 1972. 248p. bibliog. $1.25pa.

Norman has an interesting job: he writes the wine advertisements addressed to physicians by the California wine growers in medical journals. This present effort is geared to people already acquainted with wine. The 16 chapters and three appendices cover a lot of territory. He explores the basis of California wine with a sense of humor, and discusses the achievements of the California wine industry. Pointers are given for ordering wine in restaurants, and there is information on touring wineries in California and how wine is made the modern way. Some mention is made of wine and health. The work concludes with a 16-page section of his favorite wine toasts and quotes, and a self-administered wine quiz.

224. Schoonmaker, Frank, and Toma Marvell. **American Wines**. New York, Duell, Sloan and Pearce, 1941. out of print.

As a classic, and one of the first of its kind (so soon after Repeal, when the industry was flat on its back), this book should really be back in print. There is a lot of detail on the struggle of producing and picking up the vines after Prohibition and the collapse of the early twenties.

CANADA

225. Bradt, O. A. **The Grape in Ontario**. Toronto, Ontario Department of Agriculture and Food, 1972. 49p. photos. (Publication 487). free.

A basic work about hybrids grown in Ontario, illustrated with black and white photos of grapes and vines in all stages of development. The importance of this pamphlet is that it describes the work of the Horticultural Research Institute of Ontario (Vineland Station) which has been experimenting extensively since 1946 to produce French hybrids that could withstand cold and unseasonal climates. It has produced thousands of these hybrids, and is the only activity of this kind in the world. Thus, it will certainly be important if they can find a breakthrough for the *vinifera* grape.

226. Rowe, Percy. **The Wines of Canada**. New York, McGraw-Hill, 1970. 200p. illus.(col.). $5.95. LC 77-547081.

In a history that closely parallels that of the United States, immigration brought a healthy onslaught of both winemakers and wine consumers. This is a history of winemaking and wine stores, with detail on modern vineyards at Niagara, Ontario, and Okanagan, British Columbia. The technical side of quality controls is examined, as well as planting, tending, and harvesting. Material includes fruit wines (loganberries, strawberries, blueberries, and honey); sweet wines and dessert wines; and how to sell abroad. The overblown but colorful style is concerned only with the commercial manufacture of Canadian wines.

AUSTRALIA AND NEW ZEALAND

227. Australia. Wine Board. **Wine for Profit: Knowing, Selling Australian Wine**. Adelaide, 1968. 120p. illus.(part col.). maps. price not reported. LC 74-422720.

228. Australia. Wine Board. **Wine–Australia: A Guide to Australian Wine**. Melbourne, Nelson, 1968. 93p. illus.(part col.). maps. $3.95Aust. LC 72-368747.

These two very similar books share certain portions of the contents. Beginning with a factual outline of the different types of Australian wines and grapes, and wine regions, the Board turns to history, vinification practices, uses of wine, and the subject of brandy. The three "S's"–selection, storage, serving–are adequately covered, and there are glossaries and a map. The only difference here is that *Wine for Profit* is larger because it has two additional chapters for the trade–sales and services for the restaurant or bar, and the store. Also, that publication is generally free outside Australia to those who request a copy.

229. Beckwith, A. R. **The Vintner's Story**. Rev. ed. Tempe, N.S.W., Penfolds Wine Pty. Ltd., 1970. 32p. illus. diagrs. maps. free. LC 71-872869.

This pamphlet, available from Penfolds, is the promotional history of a commercial winery.

230. Benwell, W. S. **Journey to Wine in Victoria**. Melbourne, Pitman, 1960. 120p. illus. LC 61-41983.

More a tourist-type book, this work describes the wine and vinification processes in the third largest wine-producing area of Australia (after the Barossa and Hunter Valleys).

231. Cox, Harry. **The Wines of Australia**. London, Hodder and Stoughton, 1967. 192p. illus. £1.75. LC 67-111117.

A witty account (without the burning aftertaste of Monty Python) and a brief history of the major areas: Barossa Valley (South Australia), Hunter Valley (New South Wales), plus sections of Victoria and Western Australia. Cox shows, by vinification and grape types employed, how Australian wines differ from those of the rest of the world. Much detail is given here about the people involved. The appendix lists the wine growers by state.

232. Driscoll, William Patrick. **The Beginnings of the Wine Industry in the Hunter Valley**. Newcastle, N.S.W., Newcastle Public Library, 1969. 81p. illus. maps. plates. tables. (Newcastle History Monographs, 5). price not reported. LC 75-551493.

An historical document, well worth acquiring for the illustrations, which detail precisely the development of one of the largest wine-producing areas in Australia. There is much biographical information about the pioneers.

233. Evans, Len. **Cellarmaster Says: A Revised Guide to Australian Wines**. Sydney, The Bulletin, 1968. 142p. illus. $3.95Aust. LC 70-396346.

This is a publication of interest mainly to Australians, since it is a tasting guide to the many Australian wines that are, unfortunately, not available in North America. (See also entry 235.)

234. Lake, Max Emory. **Classic Wines of Australia**. Brisbane, Jacaranda; distr. by Tri-Ocean, 1966. 134p. illus. maps. $8.45. LC 67-73106.

A description of wines mainly from the Hunter Valley. Some older wines, rediscovered, are examined, but the emphasis is on the changes in Australian wines over the past 40 years.

235. Murphy, Dan F. **Australian Wine: The Complete Guide**. 4th ed. Melbourne, Sun Books; distr. by Tri-Ocean, 1971. 206p. illus.(col.). $3.35pa. LC 73-28698.

As a competition to Evans (see entry 233), this book has more applicability to North America, for it includes more wines that are available domestically. Comparing the two books, it appears that the Australians keep the best wine for themselves (the opposite of the French approach). A good book for the tourist.

236. Rankins, Bryce Crossley. **Wines and Wineries of the Barossa Valley**. Milton, Queensland, Jacaranda; distr. by Tri-Ocean, 1971. 114p. illus.(part col.). maps. tables. $7.95. LC 72-180570.

This book complements Lake's (entry 234) in describing wines of the other important areas in Australia. Full details and figures are given for the individual wine makers.

237. Roberts, Ivor Charles. **Australian Wine Pilgrimage**. Sydney, Horwitz, 1969. 162p. illus.(part col.). $5.50Aust. LC 76-455504.

A guided tour of the major wineries of Australia, with many photographs but few maps.

238. Scott, Dick. **Winemakers of New Zealand**. Auckland, Southern Cross Books, 1964. 100p. illus. $3.25; $1.75pa.N.Z. LC 66-33641.

A slim book describing the history and modern-day advancement of vinification in New Zealand, with a directory and guide to the major vintners.

239. Simon, André. **The Wines, Vineyards and Vignerons of Australia**. London, Hamlyn, 1967. 194p. illus. maps (col.). tables. £3.00. LC 68-8187.

Published in Melbourne in 1966, this is a comprehensive work by the noted wine scholar who rarely deviated from European products. His name and work now add luster to Australian wines. Part one is a geographic description, with the maps, of the vineyards. Part two is about the wines, with a general survey of the types of wines and productions (compared to Europe). Part three (almost half the book) discusses 43 vignerons by region, from their early history through the cooperative marketing schemes. The appendices, about a third of the book, contain: historical extracts, 1803-1866; Australian grape varieties and quantities produced; characteristics to look for in Australian wines; a handful of the obligatory recipes (none particularly Australian); and a description of the work of the Australian Wine Research Institute, a body comparable to Ontario's because of the inhospitable climates. The valuable bibliography (based to some extent on entry 61) is mainly historical; most references are to nineteenth century publications.

240. Thorpy, Frank. **Wine in New Zealand**. Auckland, Collins; distr. by Tri-Ocean, 1971. 199p. illus.(part col.). maps. $12.54. LC 78-28490.

More comprehensive than Scott (entry 238), this book details the history of wine and its trade in New Zealand. It is particularly good on the early struggles of the pioneers. Unfortunately, not much of this wine finds its way to North America.

SOUTH AFRICA

241. Biermann, Barrie. **Red Wine in South Africa.** Cape Town, Buren, 1971. 156p. illus.(part col.). price not reported. LC 72-183012.

This is a short account of the history of red wine vinification, with particular detail on grape types, aging, and casks. It complements De Klerk's efforts (entry 243).

242. De Bosdari, C. **Wines of the Cape**. 3rd ed. Cape Town, Buren; distr. by International Publications Service, 1971. illus. $6.25.

A more general book on a specific area (the Cape or Paarl wines), with a short history of the K.W.V. (see entry 244). Much detail here on the individual characteristics of individual grapes or wine types, plus statistics of the wine trade.

243. De Klerk, Willen Abraham. **The White Wines of South Africa**. Cape Town, Balkema; distr. by International Publications Service, 1971. 110p. illus. $6.25. LC 68-86140.

Originally published in 1967, this is a short account of the history of white wine identification, with particular detail on grape types, sweet and dry qualities, and casks. It complements Biermann's book (entry 241).

244. Ko-operatieve Wynbouwers Vernging Van Zuid-Afrika, Beperkt. **The South African Wine Industry**: **Its Growth and Development**. Paarl, South Africa, Public Relations Department of K.W.V., 1971. 1v. (unpaged). illus. price not reported. LC 72-184520.

A hodge-podge of history, pictures, anecdotes, and markets for the wines of South Africa (especially the Cape and Paarl). See also entry 383 for the K.W.V.'s publication about cookery, which contains essentially the same information.

245. Merwe Scholtz, Hendrik van der. **Wine Country**. Cape Town, Buren, 1970. 239p. illus.(part col.). 5.55Rand. LC 75-853056.

More a tour book, with lavish photographs, this book details routes and journeys along the wine growing areas and streams in the Cape area. For the specialist.

246. **Spirit of the Vine: Republic of South Africa**. Ed. by D. J. Opperman. Cape Town, Human and Rousseau, 1968. 360p. illus.(part col.). 13.50Rand. LC 68-111453.

This was published on the occasion of the Fiftieth Anniversary of the K.W.V. (Ko-Operatieve Wynbouwers Vernging). Each of the 14 chapters was written by a specialist in South Africa. With large type, large pages, and color illustrations, this book would be doomed to a coffeetable existence were it not for its excellent text. Material covered includes a description of the parent wine stock from the Near East (a brief history, but with many color illustrations showing grapes and vines) and the historical development of wine (from the Ancient World, through the Middle Ages, to the Union of South Africa by the Dutch at Cape Peninsula in 1655). The problems of growth are covered, as well as marketing strength resulting from the founding of the K.W.V.–storage of 50 million gallons, wholesale and exporter clout, vine types developed. Several chapters relate visits and descriptions of members and the estate wines. Only since 1965 has an effort been made to sell South African wines in the United States. The technical section in this book presupposes some knowledge, but the details on grafting stock and disease (with the appropriate photographs) are interesting even from a layman's point of view. Closing chapters of this comprehensive work discuss wine in art, literature, and music, with fine detail paid to the Afrikaans literature and language.

Chapter 3

BEERS

The paucity of materials on beer is not surprising when one considers that one would rather make it or drink it than read about it. Beer tastings are rare, although they do occur. Beer is a very technical product, and there are dozens of periodicals to keep the brewmaster updated on technological developments. Similarly, since much domestic beer tastes the same all over the United States (all of it lager), the lack of exposure to different types (e.g., bitters, stout, porter, bock, ale) precludes much interest on the part of the American consumer.

247. Baron, Stanley Wade. **Brewed in America: A History of Beer and Ale in the United States**. Boston, Little, Brown, 1962. 424p. illus. bibliog. $7.50. LC 62-9546.

Since beer was the universal beverage of the seventeenth century, it was natural that the art of brewing would be carried on in the Colonial period. The brewing industry, as the author notes, has played a large part in the social and cultural growth of the United States. It slowly moved from the art of the 1630s to the science of 1819 (when the first steam engine was installed in a brewery). Short biographies of the leaders in the field are given, and, of course, the impact of the introduction of German lager in the 1840s is assessed. This scholarly work details the advances made in science (pasteurization, refrigeration, transportation) and the Prohibition-Repeal period. An extensive 22-page bibliography (including unpublished material) concludes this readable book.

248. Birch, Lionel. **The Story of Beer**. 2nd ed. London, Truman, Hanbury, Buxton and Co., 1965. 96p. illus. free. LC 66-72114.

This booklet, together with its companion, *Trumans: The Brewers; The Story of Truman, Hanbury, Buxton & Co.* (London, 1966. 63p.), is well illustrated from company files. Birch dwells on what beer is all about (historical and technical) and how it is consumed, while the corporate history explains the beer-making process and the role that the corporation itself played, plus historical advertisements.

249. Birmingham, Frederic. **Falstaff's Complete Beer Book**. New York, Award Books, 1970. 151p. illus.(part col.). $1.50pa. LC 70-21037.

Produced under the auspices of Falstaff Brewery Company, this is another semi-corporate history that also deals with the world-wide availability of beer and beer products.

250. Brewers Association of Canada. **Brewing in Canada**. Ottawa, 1965. 142p. illus.(col.). free. LC 68-41444.

Beer is very important in Canada, both as an industry and as enjoyment. It has also some of the world's finest beers. This is an economist's book, with various trade statistics and many charts and figures that outline the growth of beer in Canada, consumption, tastes in beer, taxes, legislation, and technical processes. The appendix lists ten tables concerning production, consumption, sales, and historical figures. A 36-page supplement was published in 1967.

251. Hough, James S., D. E. Briggs, and R. Stevens. **Malting and Brewing Science**. London, Chapman and Hall; distr. by Barnes and Noble, 1971. 678p. illus. bibliog. $30.00. LC 75-860722.

Since it first came out, this book has become the classic technological text on beer making. Superb chapters and illustrations on malt, barley, hops, processes, pasteurization, and so forth.

252. Iatca, Michel. **Guide international de la bière**. Paris, A. Ballard, 1970. 435p. illus. $10.00. LC 79-552983.

This book, written in French, details the beer-making processes in nearly every country of the world, expending the most space on Germany, Czechoslovakia, Scandinavia, Britain, and North America. Types of beer and individual drinking habits are discussed. This worthwhile book should be translated into English for the North American market.

253. Kauffman, Donald, and June Kauffman. **The United States Brewers Guide, 1630-1864: A Complete List of American Breweries, Owners, Addresses, and Histories**. Cheyenne, Wyo., 1967. 41p. price not reported. LC 67-4079.

As comprehensive as can be, this history serves the purpose of illustrating the kinds and varieties of geographically separated brewers.

254. Kelly, William J. **Brewing in Maryland: From Colonial Times to the Present**. Baltimore, 1965. 735p. illus. price not reported. LC 66-763.

An incredibly complex work, detailed down to the last head of foam, this masterful book was produced as a labor of love. More studies of this kind are urged for corporate histories.

255. Marchant, W. T., comp. **In Praise of Ale: or, Songs, Ballads, Epigrams, and Anecdotes Relating to Beer, Malt, and Hops, with Some Curious Particulars Concerning Ale-Wives and Brewers, Drinking-Clubs and Customs**. Detroit, Singing Tree Press, 1968. 632p. $19.50. LC 68-22038.

This facsimile of the 1888 edition contains "rich, rare, and racy songs," plus everything already mentioned in the subtitle. The author dragged up many references from newspapers, friends, magazines, and books. An eclectic hodge-podge for the lover of beer.

256. Monckton, H. A. **A History of English Ale and Beer**. London, Bodley Head, 1966. 238p. illus. maps. tables. bibliog. £ 2.00. LC 66-76934.

This is a fairly complete and chronological account of the beer trade in England, and it also explains the process of malting and brewing in historical terms. It describes the rise in popularity, economic importance, and consumption of ale and beer, since the fifteenth century development from the Lowlands. Monckton explains such things as bride-ales, inn tokens, Assize of Bread and Ale in 1267, and the changing shapes of vessels. Supplementary material includes beer consumption figures for the past 300 years. Appendices list the Acts of Parliament that dealt with beer. A readable book, lightened by anecdotes.

257. Pearl, Cyril. **Beer, Glorious Beer**. Melbourne, Nelson, 1969; distr. by International Publications Service, 1971. 173p. illus. tables. $7.50. LC 70-472915.

A paean to the malt beverage, with some incidental observations on great beer myths, pubs and publicans, bar maids and breathanalyzers, poetry, and some recipes and use of beer in the kitchen.

258. Rosenthal, Eric. **Tankards and Tradition**. Cape Town, H. Timmins, 1961. 209p. illus. 1.85Rand. LC 62-736.

Relates the history and development of beer in South Africa.

259. Sigsworth, Eric M. **The Brewing Trade During the Industrial Revolution; The Case of Yorkshire**. York, St. Anthony's Press, 1967. 36p. tables. (Brothwick Papers, 31). £0.50. LC 67-87731.

A regional economic history of a prolific industry during the period of the Industrial Revolution in England.

260. Vaizey, John Ernest. **The Brewing Industry, 1886-1951: An Economic Study**. London, Pitman, 1960. bibliog. 173p. £1.75. LC 60-36340 rev.

This study was done for a British group, The Economic Research Council. It is a basic history covering 65 years of tariffs, unemployment, two wars, and changing technologies.

261. Weiss, Harry B., and Grace M. Weiss. **The Early Breweries of New Jersey**. Trenton, New Jersey Agricultural Society, 1963. 98p. illus. bibliog. $2.00. LC 63-9259.

While not as detailed as entry 254, this state history is a fine example of what can be produced for the other states in the way of a corporate history or the history of an industry.

Chapter 4

SPIRITS

The history of mankind has been so entwined with alcoholic spirits that it has been difficult to isolate those books that deal with spirits *per se*. Much material can be found, of course, as part of general works (such as texts dealing with distillation or economics, or books on such matters as prohibition, pubs, alcoholism, and so forth). The monographs in this section are primarily about the big three spirits: bourbon, cognac, and Scotch whisky. There is a noticeable dearth of material on rye, vodka, and gin.

262. Balfour, Patrick. **The Kindred Spirit: A History of Gin and of the House of Booth**. London, Newman Neame, 1959. 93p. illus. free. LC 60-31563.

A short but exceptionally readable account of the role one corporate house played in the history of gin. Gives a full explanation of the gin tax situation in the eighteenth century.

263. Barnard, Alfred. **The Whisky Distilleries of the United Kingdom**. New York, A. M. Kelley, 1969. 457p. illus. maps. $20.00. LC 69-16763.

Originally published in 1887, this reprint presents corporate histories of the distillation industry in Ireland, Scotland, England, and Wales. The maps were pretty good for their time.

264. Brown, John Hull. **Early American Beverages**. Rutland, Vt., Charles E. Tuttle, 1966. 171p. illus. bibliog. $10.00. LC 66-17771.

A detailed account of the early pioneer methods of distillation, along with a description of the end products. Drinks other than alcohol can also be found here.

265. Carson, Gerald. **The Social History of Bourbon: An Unhurried Account of Our Star-Spangled American Drink**. New York, Dodd Mead, 1963. 280p. illus. $6.95. LC 63-14884.

Utilizing vintage photographs for this history of the all-American drink, Carson details the Whiskey Rebellion, the Whiskey Ring, and the Whiskey forts of the fur trade, using both anecdotes and straight facts. American drinking manners are covered up to 1920, but only 11 pages are devoted to the past 43 years. Chapter notes are given in lieu of a bibliography, and these are mainly older accounts. Strangely enough, there is nothing here on Prohibition.

266. Crowgey, Henry G. **Kentucky Bourbon: The Early Years of Whiskey Making**. Louisville, Ky., University Press of Kentucky, 1971. 171p. illus. bibliog. $9.75. LC 79-111506.

A scholarly history of the early production of bourbon, complete with derivation of name and government controls of the initial product.

267. Daiches, David. **Scotch Whisky: Its Past and Present**. New York, Macmillan, 1970. 168p. illus.(part col.). maps (part col.). $9.95. LC 77-89930.

This is the grand tour of malt distilleries. Daiches studied the complex manufacturing process in 1967, and here gives his account, along with a social and economic history (taxes to be collected, malt and barley processes). The inventions of stills (pot versus patent) and of blending are covered. The two significant highlights in history—the amalgamation of the Big Five into the Distillers' Corporation, and the 1907 decision allowing blends—had the direct result of making Scotch the second largest United Kingdom export. Statistics, a glossary, old cartoons, and photographs by the author's son complete this compact book.

268. Hallgarten, Peter. **Liqueurs**. 2nd ed. London, Wine and Spirit Publications, 1972. 144p. illus. facsims. £1.75.

This is a concise picture of the history and manufacture of liqueurs and cordials. Each type is listed by category (e.g., herbal, sweet, distilled, macerated, etc.), and most of the world's spirits and liqueurs are covered.

269. Hicks, Wendy. **The All of Scotch Whisky**. Toronto, Distillers Company (Canada) Ltd., 1970. 46p. bibliog. free.

Sifting through much of the existing literature, Ms. Hicks has produced a cogent summary of essential information regarding its history, technical processes, blending, bottling, and selling (public relations, advertising, and anti-pollution). Interspersed throughout are the innovations and reactions of the large Distillers Company Limited (D.C.L.) conglomerate. There are brief historical notes about the Big Five–Buchanan, Dewar, Haig, Walker, and White Horse.

270. Kroll, Harry Harrison. **Bluegrass, Belles and Bourbon: A Pictorial History of Whiskey in Kentucky**. South Brunswick, N.J., A. S. Barnes, 1967. 224p. illus. $10.00. LC 67-10927.

More a social history of drinking, this book describes the production of bourbon in context with the life style of the time. Many of the illustrations are rare, having been collected from personal files.

271. Lafon, René, and Pierre Couilland. **Le Cognac, sa distillation.** 4th ed. Paris, J. B. Baillière, 1964. 270p. illus.(col.). maps. bibliog. 20 Fr.Fr. LC 67-116276.

This is the definitive book about cognac, detailing its history, the discovery of double distillation, and the effect of storage in cask. There is some description of present-day cognac houses, along with an introduction by Maurice Hennessy.

272. Layton, T. A. **Cognac and Other Brandies**. London, Harper Trade Journals, 1968. 153p. illus. £2.00.

This basic and factual description of the history and manufacture of brandy (worldwide) also includes some "must" items for the tourist in France.

273. Licensed Beverage Industries, Inc. **Moonshine: Misery for Sale**. Washington, D.C., 1973. 24p. illus. free.

This report, revised annually, is a fascinating scare document. It is not to be taken lightly, since it points out the dangers of illicit distilleries by citing the worst examples (illustrated) of "rot gut." The nation-wide survey, full of statistical information, includes data on the number of stills seized and the production and consumption of illegal spirits. For instance, the total busts in 1971 (the last year available) were 6,650 (down from a 1956 high of 25,605); Alabama led the way with 2,060, followed by Georgia and North Carolina.

274. Lockhart, Sir Robert Bruce. **Scotch: The Whisky of Scotland in Fact and Story**. 3rd ed. London, Putnam, 1966. 184p. £1.25. LC 67-88426.

As a romantic history from one who mourns the eclipse of the malt whiskies, this book describes the manufacture of Scotch from cauldron days to patent-still. Material includes a discussion of the export markets and what happened during Prohibition. Many personal reminiscences are scattered throughout (the author has known Speyside since birth). There are profiles of Dewar, Buchanan, Walker, and Haig (all real people) plus a vivid description of the conglomerate Distillers Company, Ltd.

275. McDowall, R. J. S. **The Whiskies of Scotland**. 2nd ed. London, John Murray, 1971. 167p. illus. maps. bibliog. £2.50. LC 72-873589.

The American edition, published by Abelard Schumann in 1970, came from the first British edition in 1967. This present book, which has been extensively revised, has new material and new illustrative maps. Part one details malt whisky (from barley); McDowall visited all the distilleries concerned, commenting on each and relating their histories. The book concentrates on these single malt whiskies. Part two is about grain whisky (from maize). Part three concentrates on the blended whiskies. The author visited 28 of over 2,000 different blending distilleries, and he gives a description of each as in Part one. Part four, the technical section, discusses the making of whisky, the choices of grain and the final product available for the consumer, and government controls. There are some comments about the Scotch Whisky Association, and a concluding chapter on the economics of whisky production, by I. A. Glenn.

276. McGuire, E. B. **Irish Whiskey: A History of Distilling, the Spirit Trade and Its Excise Control in Ireland**. Dublin, Gill and Macmillan; New York, Barnes and Noble, 1973. 462p. illus. bibliog. $20.50. LC 73-168382.

An economic and corporate history for the specialist, which details the extent of government controls and of competition with the Scotch whisky.

277. Misch, Robert Jay. **Quick Guide to Spirits**. New York, Doubleday, 1973. illus. $3.50.

As is usual with the Misch "quick guides," this one is crammed with facts and figures. He also dismisses myths and misconceptions as trivia. Some of the recipes here which use distilled alcohol seem to be unique. Misch is always recommended for his strong opinions.

278. Pearce, John Ed. **Nothing Better in the Market**. Louisville, Ky., Brown-Forman Distillers Corp., 1970. 96p. illus. LC 70-110212. free.

This is a well-illustrated corporate history of Brown-Forman Distillers, the manufacturers of Old Forester. Their files have produced old documents, photographs, and sketches. Also included here is a well-defined history of 100 years of spirits in America, and a description of bourbon techniques. Prohibition is examined at length, plus the counteractions taken, and marketing is also looked at. Similar books include Ben Green's *Jack Daniel's Legacy* (Rich Printing Co., 1967) and J. B. Wilson's *Spirit of Old Kentucky* (Glenmore Distilling Co., 1945).

279. Ross, James. **Whisky: The Story of Scotch Whisky from Its Beginning.** London, Routledge and Kegan Paul, 1970. 158p. illus. facsims. $6.75. LC 75-515751.

A history of Scotch, very similar to Daiches' work (entry 267). It covers the invention of stills, the discovery of blending, the extent of government involvement, and the impact of exports.

280. Scotch Whisky Association. **Scotch Whisky: Questions and Answers.** Edinburgh, 1972. 56p. free.

Here are 86 point-by-point questions and answers about Scotch whisky. Simple answers to simple questions give an excellent overview without needless detail. There is also a section on cocktails made with Scotch whisky. Subjects covered include basic definitions, history, taxes, economy, sales, techniques and processes, blending and serving.

281. Weiss, Henry B. **The History of Applejack or Apple Brandy in New Jersey, from Colonial Times to the Present.** Trenton, N.J., New Jersey Agricultural Society, 1954. 265p. illus. price not reported. LC 54-8638.

As a perfectly marvellous history about one type of spirit in one particular state, this is an exemplary book that is also a treasure trove of information. All aspects are looked into, from the apples themselves to the many forms of consumption (and where).

282. Wilson, Ross. **Scotch: The Formative Years.** London, Constable, 1970. 502p. illus. £4.00. LC 78-518534.

This economic history is largely an expanded version of the author's *Scotch Made Easy* (1960). It is a massive work chronicling the early development, the pot-still discovery, and government intervention, especially by the English.

Chapter 5

BREWING AND WINEMAKING AT HOME

The rising cost of commercial wines has encouraged winemaking at home. Handbooks, easy-to-use equipment, plastics, and good concentrates have all contributed in no small measure to the increased interest of the hobbyist. Unfortunately, the books themselves are often the source of problems: the British texts use Imperial measures and tend to produce rather sweet wines, and all texts use exact measures and exact recipes which take into account nothing concerning the balance of the final product. Tables of adjustment are needed, and these are adequately furnished by both Tayleur and Anderson, below, among others.

Only a "head" of the household can legally make (but not sell) wine (up to 200 gallons per year). This must be for "food value and medicinal purposes." The U.S. government does not produce any pamphlets on home winemaking. The amateur must register with one of the regional offices of the Alcohol and Tobacco Tax Division of the Department of Internal Revenue. He must ask for *two* copies of IRS Form 1541, "Registration for Production of Wine for Family Use," which must be filled out and returned. One copy will be retained; the other copy will be certified and returned to the registrated amateur, who must then *post* it in the wine cellar, or wherever he makes his wine. All such installations are open for inspection by federal agents. Addresses for registration offices are as follows:

California: 780 Market Street, San Francisco 94104
Georgia: 275 Peachtree Street, N.E., Atlanta 30303
Illinois: 17 North Dearborn Street, Chicago 60602
Massachusetts: 55 Tremont Street, Boston 02148
Nebraska: 215 North 17th Street, Omaha 68102
New York: 90 Church Street, New York 10007

Ohio: 222 East Central Parkway, Cincinnati 45202
Pennsylvania: 2 Penn Center Plaza, Philadelphia 19102
Texas: 1114 Commerce Street, Dallas 75202

Some states have local winemaking laws that increase the restrictions already laid down by the federal government. For example, in Utah (a dry state) no head of the household can make wine at all unless he lives on a federal military installation (i.e., on federal land). Beermaking in any quantity is illegal in the United States, but it is done. Distillation is, of course, also forbidden.

Periodicals on this subject are listed in a separate division of the "Periodicals" section of this guide (pp. 139-49) and a list of suppliers of equipment will be found in the "Directories" section.

283. Adams, John F. **An Essay on Brewing, Vintage and Distillation, Together with Selected Remedies for Hangover Melancholia; or, How to Make Booze**. Garden City, N.Y., Doubleday, 1970. 108p. illus. $0.95pa. LC 70-89138.

The author's purpose is a simple one: "to write a book about wine making and brewing based on the thesis that the person reading the book could actually do it!" Adams tells it all: what equipment is necessary, what ingredients to use, the fallacies of famous rumors about making booze, and finally, what to do about your hangover that results from reading this book. A book that is thoroughly fun—but that is also practical. Included are such hard-to-find recipes as bathtub gin and white lightning. Some of the illustrations themselves might even be illegal!

284. Adkins, Jan. **The Craft of Making Wine**. New York, Walker, 1972. 91p. illus. index. $6.95.

This lovely hand-illustrated guide to wines and wine-making would make a great Christmas gift. Charmingly done, it takes the reader on a tour of the families of wine, discusses the tools of making it, provides a step-by-step guide to a red wine and a white wine, talks of country and flower wines, and discusses concnetrates, sparkling wines, and a host of related topics. Reading a winemaker's manual is one thing, but actually being able to turn the pages and visualize each operation in pleasant and instructive drawings is another. However, this is definitely for the beginner. The information is in no great depth—just very clear shallows.

285. Anderson, Stanley F. **The Art of Making Wine**. New York, Hawthorn, 1971 (c.1970). 181p. illus. $5.95; $1.50pa. LC 72-160632.

Stanley Anderson is the inventor-operator of the famous Wine-Art chain of stores, which sell wine- and beer-making supplies, materials, and equipment. His name has become synonymous with home-brewing and home vinification. This exceptionally clear guide is a simple explanation of the principles of winemaking. Most of the common problems of the beginner are thoroughly discussed, and directions for adjusting and adpating procedures to meet local conditions are welcome additions. Anderson's comments on wine from fruits are just about the best to be found. Clear, simple, and instructive, *The Art of Making Wine* is a first buy for the beginning vintner or for the library.

286. Anderson, Stanley F., with Raymond Hull. **The Art of Making Beer**. 119p. illus. New York, Hawthorn, 1971. $5.95; $1.50pa. LC 76-169864.

As a follow-up to his popular *The Art of Making Wine*, Anderson has produced a primer for the home brewer. A clear, concise, step-by-step manual for the beginning beermaker, it lays down explicitly the principles that govern the quality of homebrew. The importance of sugar control, often overlooked in other guides, is thoroughly explored as well as all other operations (such as brewing, bottling, storage, and use of the hydrometer). The illustrations are clear and instructive and the explanation of equipment eliminates the mysteries that many other manuals don't dispell. Several dozen types of beers and ales are explained, as well as basic instructions for making cider and perry. Undoubtedly, Anderson's books on wine and beer are the best on the market–especially for those who want to buy and rely on only one "brewer's bible."

287. Aylett, Mary. **Encyclopedia of Home-Made Wines**. London, Odhams, 1957. 192p. (Modern Living Series). $3.50. LC 57-37826.

Although a shade out of date concerning new techniques and the tinned concentrates, Ms. Aylett's work is nonetheless well laid out and well organized for high reference use. Besides the basics, there are recipes for weird items like metheglin, bragget, and several other older English drinks, proving that hardiness (especially British hardiness) is needed for survival.

288. Beadle, Leigh P. **Brew It Yourself: A Complete Guide to the Brewing of Beer, Ale, Stout, and Mead**. Rev. ed. New York, Farrar, Straus, and Giroux, 1973. 109p. illus. $5.95. LC 79-164535.

Next to Stanley Anderson's *The Art of Making Beer*, this is the most useful of the home brewer's beginning books. It's all here: specific gravities, hops, campden tablets, primary fermentation, use of enzymes and yeast, beer recipes, illustrations, etc. This is basically a beer book, so don't make it a first for mead. The most important difference here is the concentration on "what can go wrong" and cures for bad beer-making habits. An excellent first-aid kit to keep next to the supply of John Bull concentrate.

289. Beadle, Leigh P. **Making Fine Wines and Liqueurs**. New York, Farrar, Straus and Giroux, 1972. 110p. illus. $4.95; $1.95pa. LC 72-84775.

Beadle presents modern-day recipes for wines, and some delightful hints for the manufacture of infusions and liqueurs (using both bought and home-made syrups). His is a more practical book than André Simon's *How to Make Wines and Cordials* (Dover Reprint, 1972. $1.50pa.), but his style of writing is a little turgid and not as enjoyable as the literate Simon book is. Thus, use Beadle for making liqueurs, but, despite it being out-of-date, read Simon for enjoyment and pleasure.

290. Berry, Cyril J. J. **Amateur Winemaker Recipes**. Andover (Hants.), England, AW Press, 1971. 124p. illus. $1.25.

Another recipe-packed home wine book from *The Amateur Winemaker*. The dozens of wines included range from the common dandelion to the exotic woodruff, from parsnip-fig to celery-apple. This one is included because of its wide variety of ingredients from which a palatable wine can be made. Non-British oenologists, though, may wish to cut the sugar recommendations in half. Try the many mulls included in this one. For beers, Berry has written *Home Brewed Beers and Stouts: A Handbook to the Brewing of Ales, Beers and Stouts at Home from Barley, Malt Extract and Dried Malt Extract* (Amateur Winemaker, 1970).

291. Berry, C. J. J. **First Steps in Winemaking: A Complete Month-by-Month Guide**. 9th ed. Andover (Hants.), England, AW Press, 1970. 149p. illus. $1.25pa.

Since its beginning in 1957 the British monthly magazine, *The Amateur Winemaker* has become one of the world's most important publications for oenologists. A spin-off of the magazine has been a series of guides, manuals, cookbooks, and hobby books for the home vintner and brewer. Of uneven quality, many of these are not worth mentioning. Others are unique in their field and deserve attention. Berry's *First Steps* has been among the most popular of AW's releases. The directions are there along with over 150 recipes, including cider and perry, mead and beer. There are also tips for judging and exhibiting. Even if the British books do have the edge on American ones in appreciation of fine wine, the oenologists have been hampered by a sweet tooth, which probably results from the lack of a plentiful supply of grapes. Two or more pounds of sugar per gallon will not yield a dry wine.

292. Bravery, H. E. **Home Brewing without Failures: How to Make Your Own Beer, Ale, Stout, and Cider**. New York, Arc Books, 1966 (c.1965). 159p. illus. index. $0.95. LC 66-17178.

Ignore the place of publication: this is British. But it is nonetheless excellent! Just add one extra pint of water to each U.S. gallon and all will be well. Bravery has produced a complete book. For the beginner, he explains simple methods, using readily prepared malt extracts and dried hops; for the more advanced brewer, grain malts, mashing, and mixing grain malts and extracts. There is also information about commercial beer-making, some delightful ruminations on cider and mead, and a few recipes that are hard to find elsewhere–treacle beer, rose-petal mead, and brown ale. All in all, a fun book, from the first fermentation to the final bottle.

293. Brown, John Hull. **Early American Beverages**. Rutland, Vt., Charles E. Tuttle, 1966. 171p. illus. bibliog. LC 66-17771.

This wonderful manual contains scads of great old-time beverage recipes–some potent and some not. All types of colonial drinks and the methods of preparing them are included. So, if you want to make posset or sassafras mead, or flummery caudle or spruce beer or even absinthe ratafia, you can do it in a traditional way. Interspersed with the recipes are bits of historical lore on American drinking habits.

294. Carey, Mary. **Step-by-Step Winemaking**. New York, Golden Press, 1973. 64p. illus. $2.50pa. LC 73-76936.

This is a very practical "how-to" book, from permit to consumption, and it is especially recommended for the complete novice. Unlike other Golden Press handbooks, this one is set in large type and large format. However, it would be even better if it were spiral bound to open flat. The first wine investigated is red wine–steps, preparing and crushing the grapes (if concentrates are not being used), adjusting the sugar content, yeast nutrients, primary fermentation, second fermentation, clarifying, acid tests, bottling and aging. There is a special section on "what to do if"–a lack of fermentation, stopped fermentation, highly chlorinated water, acidity, yeastiness, cloudiness, explosions, and syrupy textures. White wine and rosés are discussed as deviancies from the red wine section, which can be confusing to the new winemaker. Information also concerns winemaking the year round, using concentrates, fruit wines, and flower wines. Much detail on crushers, pressers, barrels, and bungs is given, in addition to a supplier list. While there are lavish illustrations on each page (and most in color), there are no recipes for other fruit or flower wines.

295. Dart, C. J., and D. A. Smith. **Woodwork for Winemakers**. Andover (Hants.), England, AW Press, 1968. 30p. illus. $1.00pa.

A clear and detailed how-to-do-it for the home winemaker who wishes to build his own equipment. Over 30 useful pieces of winemaking equipment are described, with excellent working drawings. For the vintner who has the carpentry skills to build a wine press, a fruit pulper, or a winery. Unique.

296. Delmon, P. J., and B. C. A. Turner. **Quick and Easy Winemaking from Concentrates and Fruit Juices**. New York, Hippocrene, 1973. 107p. illus. index. $4.95. LC 73-77014.

The process of making wines from juices and concentrates should be an easy one, considering the problems involved in racking and pressing raw fruit. However, the problems of artificial sweeteners, colorings, and chemical preservatives are often hard to overcome. This book offers various recipes that avoid these pitfalls. However, watch for the British measurements and the problem of availability of such exotics as Polish bilberries and Australian apricots. All in all, a commonsense approach to a sticky problem.

297. Eakin, James H., and Donald Ace. **Winemaking as a Hobby**. University Park, Pennsylvania State University, College of Agriculture, 1970. 63p. illus. $2.50pa.

At a rock-bottom bargain price, this oversize (8½ x 11") work of art comes complete with well-drawn and well-designed diagrams and charts. The writing style is remarkably lucid for two people who are basically explaining how *they* run their hobbies. It compares favorably with Carey's book, although it lacks the latter's illustrative details.

298. Ehle, John. **The Cheeses and Wines of England and France: With Notes on Irish Whiskey**. New York, Harper and Row, 1972. 418p. illus. $10.95. LC 72-769659.

This interesting but very practical book is notable for coming up with a complete "wine and cheese" party from scratch. Ehle relates the history and technique of making wine *and* cheeses at home, and it is quite apparent that he has had much practical experience. Of note are the older recipes for cider, perry, and mead, in addition to French red and white wines, champagnes, whiskey, and English ale. There are recipes for about 50 varieties of cheeses, both hard and semi-soft. This is a very readable book, since the recipes are not continuous but rather are scattered among the anecdotes and personal reminiscences of the author's travels. For the serious student who also wishes to produce his own cheeses.

299. Fessler, Julius. **Guidelines to Practical Winemaking**. Oakland, Calif., 1965. 96p. illus. $3.75pa. LC 68-29612.

Combining a learned discussion on wines with wine-making, Mr. Fessler, a contributor to *Wines and Vines,* sets out to straighten the record of what home-made wine is and is not. In doing so, he presents a unique section on making wine vinegar (it does not always appear by accident). This book is available from the author at Box 2842, Rockridge Station, Oakland, Calif. 94618.

300. **The Foxfire Book: Hog Dressing, Log Cabin Building, Mountain Crafts and Foods, Planting by the Signs, Snake Lore, Hunting Tales, Faith Healing, Moonshining, and Other Affairs of Plain Living**. Ed. by Eliot Wigginton. New York, Doubleday, 1972. 384p. illus. index. $8.95. LC 70-163087.

Drawn from *Foxfire Magazine*, these tape recorded and photographically illustrated pieces of old-time lore are both instructive and fascinating. None is more fascinating than the chapter, "Moonshining as a Fine Art." This extensive article gives detailed instructions on distilling, illustrates what it says with clear and understandable drawings, and provides a glossary of the expressions and terms of "stilling." If you want to risk making 195-proof whiskey ("High Shots" to the initiated), this book is a must.

301. Gennery-Taylor, Mrs., pseud. **Easy to Make Wine: With Additional Recipes for Cocktails, Cider, Beer, Fruit Syrups and Herb Teas**. New York, Gramercy, 1957. 124p. illus. index. out of print.

The key to this slim guide is simplicity/economy. Originally published in England as *Easymade Wine and Country Drinks* (and last revised in 1959), this is one of the few vintner aids that ignores fancy equipment and recommends winemaking in saucepans, preserving pans, earthenware bowls, and bottles and corks. It's really a countrywoman's guide, with delightful statements such as "cowslip wine will cure jaundice." The information on how to make wine, beer, cider, mead, cordials and syrups is as clear and simple as the equipment needed. The added material is welcome: wines as remedies, Christmas drinks, the wine calendar, and the herb tea and dandelion coffee recipes. The wine recipes themselves are in the "peasant wine" tradition–clove and carrot, marigold and tomato, wheat and barley. It's a good one, but remember to add to the gallon–it's British!

302. Geoffrey, John. **Modern Methods of Cider Making**. National Association of Cider Makers, 1970. 25p. illus. free.

This handy little manual is an excellent guide to producing cider, the fermented fruit juice of the apple. If you live in an area where grapes won't grow, then cider-making is one way to exercise that personal satisfaction that comes from making a mildly intoxicating drink from beginning to end. This guide gives directions for selecting the fruit, crushing, pressing, removing solids, and preventing oxidation (the scourge of apple and pear juice). The brewing directions are simple and concise and the storage information seems safe enough.

303. Hardwick, Homer. **Winemaking at Home**. Rev. ed. New York, Funk and Wagnalls, 1970. 258p. illus. index. $6.95. LC 68-13032.

In an excellent article in the back-to-the-land publication *Mother Earth News*, Edward A. Wilson, the "Little Old Oberlin Wine Maker" (No. 5, pp. 54-57) places Hardwick's book at the top of the list for the best all-purpose winemaking book award. He is probably right. This is *the* complete vintner's manual. Its real strength lies in preventing the amateur from spoiling his wine through unsanitary conditions, microorganisms, or temperature fluctuations. It does not imply that winemaking is simple enough for anyone to attempt, and in dealing with the possible problems Hardwick shows the reader the way to fine wine. Well illustrated and capably written (with over 200 recipes), this is the long-awaited revision of the 1954 original edition.

304. Hutchinson, Peggy. **Home Made Wine Secrets**. New York, Drake, 1972. 124p. $4.95. LC 79-17597.

An excellent up-to-date amateur manual for winemakers which does take into account the explosion of interest in oenology and the subsequent flood of available winemaking products. A rather folksy approach combines with simple instructions for a blend of enjoyable reading and understandable directions. The first wine manual with a really clear explanation of how to unstick a stuck fermentation, but unfortunately, it has *no* illustrations.

305. Jagendorf, Moritz. **Folk Wines, Cordials and Brandies: Ways to Make Them Together with Some Lore, Reminiscences, and Wise Advice for Enjoying Them**. New York, Vanguard, 1963. 414p. illus. $12.50. LC 63-21854.

This sage approach to the folk art of creating one's own intoxicants is thought of as the best of the country wine manuals. Jagendorf intersperses his instructions with meandering memories, wisdom tempered with the humor of mistakes, and gentle advice on the use and abuse of what this book is about—the creation of excellent wines and other spirits. Buried in the absorbing narrative is much information on country wines, musts, aging, seasons for fermenting, flavors, and fruits. A thoroughly charming reading experience.

306. MacGregor, D. R., J. A. Ruck, and J. F. Bowen. **Home Preparation of Juices, Wines, and Cider**. Ottawa, Canada, Department of Agriculture, 1970. 22p. illus.(col.). (Publication 1406). free.

A helpful guide for the beginner, this Canadian government document, while written in typical documentese, is a rather useful and instructive manual. The directions are in Imperial gallons and, like its British counterparts, its recommendations for the use of sugar cater to the sweeter tastes of the British and Canadians. A handy publication for those just thinking about home brewing.

307. Milligan, Barbara J., ed. **Wine-Art Recipe Booklet**. Vancouver, Wine Art, 1971. 62p. illus. $1.00pa.

Stanley Anderson, founder of Wine-Art, a leading manufacturer of wine- and beer-making supplies, also sees to it that his franchises carry a supply of Wine-Art wine booklets. These are particularly good if the winemaker is using prepackaged concentrates (Wine-Art or some other brand). The directions and recipes are keyed to the tins of concentrate and packages of acid blend and enzyme powder and so forth. There are also recipes for wines from natural fruits and grapes. The introductory material includes an intelligent discussion of the "Pearson Square," one method of determining the amount of alcohol needed to fortify a wine. Also edited by Milligan is Wine-Art's excellent 39-cent, 20-page booklet, *Brother Vinheart's Country Wines*. It is similar to the above, except for a heavier emphasis on fruit and flower wines.

308. Orton, Vrest. **The American Cider Book: The Story of America's Natural Beverage**. New York, Farrar, Straus, and Giroux, 1973. 136p. illus. $6.95. LC 72-97610.

Orton has written widely on the subject of Vermont, and it is only natural that he would proceed to that state's natural drink made from apples. A short history of cider is presented (including both legends and American usage), along with general methods of amateur and commercial cider-making (using both old-time and modern principles). The production of cider apples is noted; there are copious illustrations of cider presses (*and* the names and addresses of manufacturers of such presses); there is a short discourse on fermenting sweet cider into hard cider; and, of course, simple directions on how to make cider at home. This book is completed by an extensive beverage and cooking section: 34 historic recipes for punches, nogs, and nightcaps (1745-1960), along with 12 modern ones, and 59 recipes for food, including dried apple pie with cider, wild game basted with cider, apple bread, and an unbelievable 16 recipes on the uses of boiled-cider pie. Although the entire book is mouth-watering, it warns the user that the results are not guaranteed.

309. Shales, Ken. **Brewing Better Beers**. Andover (Hants.), England, AW Press, 1970. 78p. illus. $1.25pa.

British beer and ale—a delectable thought! *The Amateur Winemaker*, Britain's monthly magazine for oenologists, has also published a series of useful brewer's books. Shales's work is a lively paperback with many recipes for all types of malt liquors from light lagers to the blackest double stout. This is really an addition to the advanced brewer's library. Dean Jones's *Home Brewing Simplified* (15p.) is for the beginner. Ten basic beers are described, with detailed directions that will succeed if followed with patience and care. C. J. J. Berry's *Hints on Home Brewing* is a more concise "rapid course" for home beer-makers, with basic essentials clearly spelled out. British brewing and winemaking guides are often confusing in directions, measurements, and terminology; but at these prices, they are welcome additions to the literature.

310. Slater, Leslie G. **The Secrets of Making Wine from Fruits and Berries.** Lilliwaup, Wash., Terry Publ. Co., 1965. 90p. illus. $1.00pa. LC 65-1858.

Another delightful vintner's book that is not well known. An excellent commonsense approach to unlocking the technical mysteries of country winemaking. The emphasis is on what you can grow and find in the way of berries and fruits–some older concepts not found in other wine manuals. Excellent material on fermentation processes and timing, fruit molds, fruit flies, etc. Includes all of the common fruits and berries plus some not so common.

311. Steinlage, Gerald F. **Wines, Brewing, Distillation**. Carthagena, Ohio, Messenger Press, 1972. 91p. $3.95. LC 72-189987.

A unique book that will attract little notice is Steinlage's *Wines, Brewing, Distillation*. The basic principles of winemaking, brewing, and the distillation of spirits are clearly outlined, along with the problems and solutions for each type of potable. Especially good on home use of hydrometers and the yeast process. The types of yeast and their effect on wine quality is given consideration in this informative and useful guide.

312. Tayleur, W. H. T. **The Penguin Book of Home Brewing and Wine Making.** Baltimore, Md., Penguin Books, 1973. 336p. illus. $2.45pa.

Although it contains few illustrations and no bibliography, this book has to be the most comprehensive around. Tayleur, the author of many brewing histories, runs a professional advisory service for amateurs. The entire book is very technical and is recommended for the serious student. Copious explanatory notes cover, in detail, fermentation, yeast, basic ingredients, equipment and processes. He goes into the re-use of yeast (a topic other books ignore), filtering, and clarifying (always tricky). Section Two covers brewing: history, uses of hops and malt, kits, and–very important–the *formulation* of recipes. Examples here include the rare Brapple and Rowan Ale. Section Three (the largest) covers wines: history, kits, concentrates versus grapes, wines of fruit and flower (dried, berries, herbs, cereal, root, vegetables, leaves, tree saps), and sparkling wines. It concludes with a superb and very important discussion on *blending*. Section Four covers fortified wines and liqueurs, including port, and a clear discussion of the Pearson Square for calculating the addition of alcohol. Section Five is cider and perry; Section Six is mead (honey). There are conversion tables and a well-developed index to each section so that there will be no confusion as to what is being looked up. The exceptional glossary has a unique concept: at the beginning of each definition, the word is identified as pertaining to either brewing or winemaking. Everything possible is here except distillation; truly a masterpiece of an instruction manual.

313. Taylor, Walter S., and Richard P. Vine. **Home Winemaker's Handbook**. New York, Harper and Row, 1968. 195p. $5.95; $0.95pa. (from Award Books). LC 67-28835.

For the eastern wine drinker, the name Taylor is quite familiar. This Taylor is the son of the New York State wine family that grows all of those grapes in the western part of the state. Taylor, who now operates his own experimental winery, covers much the same ground as do other well-known winemaking manuals. However, the fascinating section on growing grapes for personal uses is well worth the price of the book. This work is extremely useful for the thoughtful American home winery.

314. Tritton, S. M. **Tritton's Guide to Better Wine and Beer Making for Beginners**. New York, Dover, 1970 (c.1969). 153p. illus. $1.75pa. LC 76-99104.

Originally published in England (Faber and Faber, 1965), this Dover reprint is a gift to vintners wherever they practice their art. The complete wine- and beer-making book, it contains a great deal of information packed into a few pages. The chapter on racking, stabilization, clarification, and fining of wine is intelligent and time-saving. Particularly interesting is the introductory chapter for beginners, in which the author recommends starting out making three simple wines (orange, apricot, and apple)–a sensible approach, since grapes can prove difficult to handle for the novice. Also included is material on aperitifs, liqueurs, showing and judging of wines, and wine faults, with preventions and cures. The section "Wine Recipes from A to Z" includes many not found readily anywhere else. Ms. Tritton has written many other such books, but this one is her best.

315. Turner, Bernard C. A., and C. J. J. Berry. **The Winemaker's Companion: A Handbook for Those Who Make Wine at Home**. 2nd ed. London, Mills and Boon, 1965. 228p. illus. £2.00. NUC 66-67017.

This book is very similar to Taylor's and to Tritton's. In fact, Turner ranks among the top of the professional amateurs in England, along with Bravery, Berry (his co-author here), Beadle, and Tritton. Stanley Anderson is his North American counterpart. In choosing books of recipes and so forth, it is perhaps wise to stick to one author (as with selecting cookbooks). Turner has also written *The A-B-Z of Winemaking* (Pelham, 1966); *Improving Your Winemaking* (Taplinger, 1967); and *A Practical Guide to Winemaking* (Hutchinson, 1966).

316. U.S. Department of Agriculture. **Growing American Bunch Grapes**. Washington, D.C., Government Printing Office; distr. by SuDocs, 1970. 21p. illus. (Farmers' Bulletin No. 2123). $0.15.

A straightforward, indicative title, describing the best ways of growing grapes in the United States. Though the emphasis is on table grapes, some pointers are useful for the *vinifera* variety (wine grapes) as well.

317. Wagner, Philip M. **American Wines and Wine Making**. New York, Knopf, 1956. 264p. illus. $6.95. LC 56-5436.

This is a classic, practical manual for the small producer and home winemaker. First published in 1933 as *American Wines and How to Make Them* to help the amateur during Prohibition, and revised for the second time in 1956, it is in need of slight revision at this date. Wagner emphasizes the principles of wine-making, and believes that America can produce wines of great quality (as shown by the historical chapters here on American viticulture). Chapters on French, New York, and California wines and winegrowing all contribute to his practical use of wines. Exact detail is presented on equipment and layout design for winemaking, supplemented by charts and excellent illustrations–unfortunatley not in color. All of this is held together by a superb 80-page introduction to the background of what wines are all about. This is for the home winemaker with some experience in the vinification of grape varieties, with little on other fruit and flower wines.

318. Wagner, Philip M. **A Wine-Grower's Guide**. 2nd ed. New York, Knopf, 1965. 224p. illus. $6.95. LC 65-18752.

Wagner, owner of Boordy Vineyards, here emphasizes grape-growing rather than winemaking. He presents a detailed description of the many varieties available for growth in North America (both *labrusco* and *vinifera* strains), their common and uncommon hybrids, and their methods of cultivation. This book is more for the ultimate hobbyist who prefers to grow his own grapes.

Chapter 6

AUDIOVISUAL SOURCES OF INFORMATION

Alcoholic beverages lend themselves well to visual appreciation. Many wine books are illustrated with colored pictures of grape clusters, wine labels, the vinification process, the vineyards, and so forth. One (Johnson's *Wine Atlas*) is even devoted to maps and topographical layout. In this section are detailed lists of films, filmstrips, slides, maps, records, tapes, kits, posters and prints, and even a radio program. Some of these are available for private purposes; others can be rented or borrowed only for group viewings. In any case, many should be available through the local public library.

FILMS

WINE

319. **Beaujolais, premier fleuve de France**. French Tourist Office. 21 min. sound, black and white. 16mm.

The story of the wine country where each man is his own wine expert, with commentary on the cultural life. Economics and social events are intimately attached to the vineyards.

320. **Daily Double**. Wine Advisory Board; distr. Association-Sterling Films (U.S.A.). 20 min. sound, black and white. 16mm.

A training film for waiters and waitresses, emphasizing practical hints for suggesting wines, selling and serving wines, increasing profits and tips, along with humorous touches. The leaflet "Index of Wine Types" accompanies the film for distribution to the audience. Free rental, or for sale by the Wine Advisory Board.

321. **The Day Grandma Came to Call**. Mogen David Corp., 1962; distr. Committee on International Non-Theatrical Events (U.S.A.). 23 min. sound, color. 16mm. LC FiA 63-284.

Highly dramatic tale of a grandmother who, after hearing Mogen David's current advertising slogan, "Wines like grandma used to make," tours the wineries, sees the facilities, and is impressed by the size of the Mogen David plant and the carefulness with which the wine is prepared.

322. **Getting Down to Earth**. Wine Advisory Board, 1964; distr. Association-Sterling Films (U.S.A.). 20 min. sound, color. 16mm. LC FiA 64-603.

An industrial film for training restaurant personnel; emphasizes how easy wine is to suggest, sell, and serve (especially California wine). Humorous scenes for audience involvement, but essentially for profit-making restaurants. Free rental, or for sale from the Wine Advisory Board.

323. **Make It an Occasion**. Gold Seal Vineyards, 1971; distr. AFP Distributors. 23 min. sound, color. 16mm. or standard 8mm. LC 78-715464.

Shows various methods of developing a vineyard, growing different varieties of grapes, and harvesting the fruit. Demonstrates the making, storing, aging, and bottling of wine and presents the different procedures for making champagne, sparkling burgundy, and cold duck.

324. **Of Time and the Vintner**. Wine Advisory Board, 1961; distr. Association-Sterling Films (U.S.A.). 30 min. sound, color. 16mm. LC FiA 62-1566.

This film follows the seasonal cycle of wine preparation, viewing the cultivation of the grapes and processes which include aging in both wooden vats and glass bottles. It also includes suggestions for the use of wine with a variety of foods. Free rental, or for sale from the Wine Advisory Board. An earlier version was made in 1957 (27 min.).

325. **Royal Purple**. Monarch Wine Co., 1961; distr. Association-Sterling Films (U.S.A.). 25 min. sound, color. 16mm. LC FiA 62-1568.

With background music by Skitch Henderson, this mildly entertaining item presents the history of wine, discussing both domestic and foreign vintages along with variations of their color, effervescence, aging, and serving.

326. **Test Your Taste**. Wine Advisory Board, 1958; distr. Association-Sterling Films (U.S.A.). 5 min. sound, color. 16mm.

Explains how to play a wine-tasting game with California wines, and points out that sight, smell, and taste contribute to the enjoyment of wine. Free rental, or for sale from the Wine Advisory Board.

327. **Vines in the Sun**. KPIX, 1964. 30 min. sound, black and white. 16mm. LC FiA 65-498.

For the series "San Francisco Pageant," KPIX got Burgess Meredith to narrate still photographs selected from old photos and glass photo plates to tell the story of winemaking in California.

328. **Wine Country–U.S.A.** Taylor Wine Co., 1964. 27 min. sound, color. 16mm. LC FiA 65-446.

This shows the place of wine in American life by following the members of a photo-journalism team as they make a study of the wine industry in the Finger Lakes district of New York. It describes the culture of the vineyards, winemaking, and the use and service of wine.

329. **Wine Growing in America**. Wine Advisory Board, 1940; distr. Association-Sterling Films (U.S.A.). 15 min. sound, black and white. 16mm. LC FiA 64-600.

Discusses the importance of the wine industry in the United States just after Repeal. Describes the growing of grapes and the production and serving of wine.

330. **Wine Industry in Portugal**. Walt Disney Productions, 1965; distr. International Communications Films. 3 min. silent, color. 8mm. (standard and super). LC 70-703783.

This segment, taken from the 1965 film *Portugal*, comes as a loop film mounted in cartridge, with a teacher's guide. It illustrates the steps involved in the process of winemaking in Portugal, and points out that wine is one of Portugal's principal exports.

331. **Winemakers in France**. Institut für Film und Bild, Munich, 1969; distr. Films Inc. (U.S.A.). 15 min. sound, color. 16mm. LC 70-705473.

As one of the "Man and His World" series, this film shows how mass production has replaced traditional methods in the French wine industry. Follows the winemaking process from pruning the vines, picking the grapes, and crushing the grapes to storage in casks or hand bottling.

332. **Women's Lib–French Style**. Comité Interprofessionnel du Vin de Champagne, 1972; distr. Universal Pictures (U.S.A.). 13 min. sound, color. 16mm.

An unusual film, in that the social, cultural, and athletic changes that have affected women in France are interwoven with the history of champagne wine.

333. **The Wonderful World of Wine**. Wine Advisory Board; distr. Association-Sterling Films (U.S.A.). 20 min. sound, color. 16mm.

A general film about wine lore and wines of California, accompanied by leaflets of the same title for distribution to the audience. One of the best films to show at clubs. Free rental, or for sale by the Wine Advisory Board.

BEERS

334. **The First 8,000 Years.** Anheuser-Busch, 1970; distr. Modern Talking Pictures Service (U.S.A.). 20 min. sound, color. 16mm. LC 70-710960.

This film traces the 8,000 years of recorded history of the brewing of beer.

335. **The Fifth Ingredient**. Molson's Brewery, Montreal, 1964 [1959]; distr. National Film Board of Canada. 22 min. sound, color. 16mm. LC FiA 64-910.

This is a revised version of the 31-minute 1959 film of the same title. It traces the history of Molson's Brewery, established in 1786, and then shows how lager and ale are made at the Brewery.

SPIRITS

336. **Kindred Spirit**. Distr. Scotch Whisky Association (U.K.). 9 min. sound, color. 16mm.

This animated cartoon provides a lighthearted look at the problem of good international relations and spirits. Free rental, but write for details of transatlantic shipping.

337. **Time Was the Beginning**. Distr. Scotch Whisky Association (U.K.). 19 min. sound, color. 16mm.

This is a general documentary film of the Scotch whisky industry, showing the complete process from birth of the grain to consumption. It is available with commentaries in English, French, German, Dutch, Italian, Spanish, and Japanese. Free rental, but write for details of transatlantic shipping.

338. **A Wee Dram**. Distr. Scotch Whisky Association (U.K.). 24 min. sound, color. 16mm.

The distilling of Scotch as seen through the eyes of a Canadian visitor. Free rental, but write for details of transatlantic shipping.

FILMSTRIPS

339. **California Wine Country**. Herbert E. Budek Films and Slides of California, 1970. 31 frames. color. 35mm. LC 76-736098.

Explains the steps in winemaking, including planting, cultivation, and aging processes. Outlines the history of the California wineries, with captions.

340. **Grocery Sales Pay Off**. Falstaff Brewing Corp., 1959. 88 frames. black and white. 35mm. LC Fi66-1577.

A commercial filmstrip, demonstrating how beer distributors can encourage grocers to stock their product by explaining that beer increases sales, resulting in faster turn-over and greater profit. It includes discussion questions, but it is of little interest to a general audience.

341. **Smooth Sale-ing**. Falstaff Brewing Corp., 1959. 75 frames. black and white. 35mm. LC 66-1612.

This interesting history of Falstaff emphasizes its product and its methods of advertising, through the use of photographs and drawings.

342. **The Wine Harvest of Moselland, Germany**. Herbert E. Budek Films and Slides of California, 1963. 33 frames. color. 35mm. LC 78-735849.

After giving the history, geographical and physical features of the Moselle area of Germany, the strip describes wine cultivation and harvesting from the gathering of grapes in the hills to the filling and labelling of bottles and shipping. With captions.

SLIDES

Two valuable slide sets are available for purchase:

343. **German Wines**. German Wine Information Bureau, 1973. $10.00. 50 slides. color. 35mm.

All facets of the German wine industry, from grape to consumption. Includes text.

344. **Scotch Whisky**. Scotch Whisky Association, 1970. £2.00. 20 slides. color. 35mm.

Illustrates the distilling and blending process. Includes text.

MAPS

Maps are found in many of the related books, and some of them (such as those in Johnson's *World Atlas of Wine*, entry 19) are in great detail. But most travelling or road maps are available only from cooperatives or well-organized tourist offices. And they are usually free. All the French maps—enough to cover your living-room walls—can be obtained just by dropping a note to the supplier—a Comité (see p.155). The following two maps, unfortunately, must be paid for:

345. **Harpers Distillery Map of the Scotch Distilleries**. 3rd ed. Harper Trade Journals (Southbank House, Black Prince Road, London, SE 1). $2.00 post free.

This colorful, large map (48" x 48") shows the exact locations of all Scotch whisky distilleries, the roads, and the airports.

346. **Wine 'n' Where**. Caligraphics, Yountville, Calif. 94599. $0.57 ppd.

A huge map (48" x 64") that is a guide to the Napa-Sonoma-Mendocino wine area of California, with sketches of the wine industry, recipes, and local lore.

PHONORECORDS AND TAPES

An exhaustive search has turned up only the following items, which we would rate as "fair to good."

347. Lichine, Alexis. **The Joy of Wine** [phonorecord]. MGM E 4519-2 (2 discs). $9.99.

This is an entertaining guide for the wines of Europe and the United States, a one and a half hour musical tour of vineyards of France, Germany, Italy, Spain, and California. At breakneck pace, he examines wines from the viewpoint of their historical backgrounds, regional characteristics, and other individual nuances. Other material concerns the service of wine and wines that complement food. He interviews Baron Philippe de Rothschild, Claude Taittinger, Count de Chandon-Moët, and lesser luminaries. The packaging includes vintage charts, maps, a glossary of wines and their phonetics, plus his own classification table of the red wines of Bordeaux.

348. **The Little Wine Uncomplicator** [cassette]. Things on Tape, 4714 N.E. 50th Street, Seattle, Wash. 98105. $5.95 ppd.

Designed for the novice, this 90-minute tape contains basic wine information: historic notes, a simple explanation of fermentation, basic classification of wines, and suggested wine tasting procedures.

349. **Menu French and Wine Labels** [cassette]. Wine Guild, Box 851, New York, N.Y. 10010. $6.50 ppd.

A basic pronunciation guide to 150 leading estate wines and 100 popular French dishes. Detailed reference sheets are included.

350. **Wine Making at Home** [cassette]. 30 min. Semplex of USA, Box 12276, Minneapolis, Minn. 55419. $3.50 ppd.

Needless to say, there is no great push on exploiting home winemaking because of its marginal legality. Thus, there is little, if any, graphic illustration beyond a few books, periodicals, and this tape. A soft, soothing voice describes in detail how to make wine at home.

351. **Wine Pronunciation Guide** [cassette or reel-to-reel tape]. Wine Tapes, Box 510, Corte Madera, Calif. 94925. $12.50 ppd.

For one and a half hours, unidentified wine merchant linguists read over 1,800 names of French chateaux, vineyards, and areas, with terms in French, German, Italian, Spanish, and Portuguese.

KITS

352. **Victor's Pocket Wine Guide**. 1973– . Vintage Associates, 3241 Norcross, Dallas, Texas 75229. Annual. $2.95.

This is a collection of a number of 3" x 6" cards inside a laminated cover. The obvious intent is to have a "ready reference" source close at hand, particularly when shopping for wines or selecting a wine in a restaurant. Crammed into this handy little package is a wealth of information, mostly available in other sources but only in book form: vintage ratings for the past 13 years; an estimate of how nearly ready to drink are the wines from a particular region; an indication of the long-lived wines; recommendations for food and wine combinations; suggested serving temperatures, as well as the timing for opening; recommendations of optimum vintages from previous years to drink *now*; a listing of the best-known wines of each region, with their price ranges; comments on the latest European vintages; some discussion on California wines; and general information on decanting, glasses, and corkscrews. The whole package was put together by Victor B. Wdowiak, Director of the Winecellars at Neiman-Marcus, and it will be revised annually.

353. **The Way of Wine: An Indispensable Aid to the Novice Hosting a Wine Tasting**. Magok, Inc., Box 27, Jamestown, Mich. 49427. $6.60 ppd.

This boxed kit has been billed as a "group educational program for the acquisition of knowledge about wine." Each box contains a 52-page booklet describing in broad detail the wines of the world, and adding suggestions on selecting wines for a wine tasting. Other material includes comprehensive charts listing over 800 wines, coded by price and classified for easy selection by country, region and type. "Invitations to a wine tasting" are included, as well as score cards, which are filled out during sampling. Ten aspects of wine evaluation and their corresponding descriptive terms are given, with easy reference to the charts. One box is all that is needed: the host can easily make his own invitations and score cards for subsequent parties.

RADIO PROGRAMS

Thus far, we have been able to find only one radio program devoted solely to alcoholic beverages. This is "The Topic Is Wine," on New York's WQXR, Monday to Friday at 7:05 p.m., with Terry Robards.

POSTERS, PRINTS, AND SOUVENIRS

For current posters, prints, ashtrays, and "bric-a-bracs," it is best to write to the beverage producer concerned. Most will send labels, maybe even an occasional cork or other sealer, or a bottle. It has been our experience that such items can be more easily obtained from foreign producers when visiting their offices abroad. These "souvenirs" are often handed out to tourists rather than to natives. However, some are not obtainable any more because they are no longer being produced (e.g., Cinzano glazed ash trays). For rare posters and prints, Elizabeth Woodburn of Booknoll Farm (Hopewell, N.J. 08525) stocks a wide variety depicting scenes of grape clusters, wine chateaux, champagnes, various aperitifs (e.g., absinthe, dubonnet, etc.), tavern drinking, and other lithographs and engravings related to consumption of beverages.

Chapter 7

ANCILLARY ASPECTS OF WINES, BEERS, AND SPIRITS

COCKTAILS

For those who do not drink neat forms of alcoholic beverages, here is a selective list of books on cocktail preparation (including wine punches, beer drinks, egg nogs, etc.). The arrangement of most of these books is, first, by the type of beverage, and then alphabetical by title of the drink. Some non-alcoholic mixes are also given so that abstainers will not feel left out of the party.

354. Cotton, Leo, ed. **Old Mr. Boston: Deluxe Official Bartender's Guide; A Collection of Recipes for Mixed Drinks to Suit Every Taste and Occasion**. Boston, Mr. Boston Distiller, 1960. 149p. illus. free. LC 60-15939.

First published in 1935 but unrevised since 1960, this collection of 1,400 cocktail preparations is strong on egg nogs and martinis. The illustrations feature Mr. Boston products (used in the recipes). But the price is right, and additional material includes bar hints, suggestions, and a liquor dictionary.

355. Duffy, Patrick G., and James A. Beard. **The Standard Bartender's Guide**. New York, Pocket Books, 1971. 254p. index. $0.95pa.

Lists 1,200 mixed drinks, including some old passé drinks such as sangarees, shrubs, etc. Additional material includes a Frank Schoonmaker vintage chart for wines by region, information on buying and serving wines, glossaries, measurements, charts of food and wine complements, dispensing draught beer and recipes for hors d'oeuvres, sandwiches, and food for large parties.

356. Mario, Thomas. **Playboy Host and Bar Book**. Rev. ed. Chicago, Playboy, 1971. 342p. illus. $14.95.

Mario, a consultant for the industry and Food and Drink Editor of *Playboy* for over 20 years, has put together a lavish, oversized book that is a guide for party-giving, with over 800 mixed drink preparations. Additional material includes background information on spirits, wine and beer, comprehensive liquor and garnish lists, suggestions for starting a wine cellar, a glossary of wine terminology, a description of bar equipment, liquor anecdotes and history, a short course on mixology, a chapter of hors d'oeuvres recipes and suggestions for 15 parties (from brunch to wine tasting to an urban luau). The color photographs are stunning.

357. Powell, Fred, comp. **Bartender's Standard Manual**. San Angelo, Tex., Educator Books, 1971. 107p. $6.95. LC 72-174494.

With the aim of standardizing drink recipes (so that "each martini in New York City" can be "equally perfect"), this little book lists over 700 recipes alphabetically by name. There is a supplementary 12-page section of nogs and party punches. Because there is no index, recipes can be located only under one name, which is a distinct handicap. In addition, an arrangement or an indexing by type of liquor (bourbon drinks, gin drinks, etc.) would be helpful. As it is, the manual will certainly be useful to actual bartenders—for whom, after all, it was intended.

358. Simmons, Matty. **The New Diners Club Drink Book**. Rev. and enlarged. New York, New American Library, 1969. 287p. illus. $0.95pa. LC 79-9353.

As the direct competitor to Duffy, this item offers virtually the same amount of information. To recommend one over the other would be a matter of tossing a coin.

359. **Trader Vic's Bartender Guide**. Ed. by Victor J. Bergeron. Rev. and exp. ed. Garden City, N.Y., Doubleday, 1972. 442p. illus. index. $6.95. LC 72-76212.

Originally published in 1947. Its revision makes it the newest of the cocktail preparation books. Over 1,000 preparations include 143 originals as served in the T.V. restaurants. There are no variations (just one standard drink per mix), and the passé drinks have been dropped. Can be used as a guide for professional bartenders.

COOKERY AND FOOD

Wine demands food, for it is the ultimate digestive beverage. At the same time, there are many cooking methods that utilize the distinct flavor enhancement of wines, beers and spirits. Out of literally thousands of cookbooks, here is a highly selective list of those that concentrate on good food with alcoholic beverage flavor, and on alcoholic beverage flavor in good food. In some cases, books in other sections will also contain recipes (see Bain, entry 97), just as cookbooks will often devote a chapter to wines or spirits (e.g., Balzer, entry 361).

360. American Heritage. **The American Heritage Cookbook and Illustrated History of American Eating and Drinking**. New York, American Heritage Publ. Co.; distr. New York, Simon and Schuster, 1964. 640p. illus.(part col.). $15.00. LC 64-21278.

Part One, the historical section, is in narrative prose and includes anecdotes and superb, lavish illustrations. Each chapter is written by a different person, with character sketches of leading historical characters (e.g., Mark Twain, Diamond Jim Brady, Delmonico's, etc.). Progression is from the Indians, through the Colonial period to the Deep South, cosmopolitan (i.e., European and Asian) tastes, and chapters on eating out and at home. Part Two is the recipe section, which gives 500 historical and unusual tested recipes, plus 30 recreated menus from the past. Unlike European cookery, American cookery seems to concentrate on the culinary uses of beer and bourbon. For a history of European eating and drinking, see the companion volume to this one (entry 378).

361. Balzer, Robert L. **Adventures in Wine: Legends, History, Recipes**. Los Angeles, Ward Ritchie, 1969. 114p. illus.(col.). $7.95. LC 76-88721.

This is primarily a cookbook (with photographs by the author), although there are extensive sections on the development of taste in wine, vineyards around the world, anecdotes, and history. In addition to wine, it covers sherry, madeira, vermouth and champagne, and these are reflected in the recipe section as well. There are recipes for all courses.

362. Bottles and Bins. **Recipes**. Francis L. Gould, ed. St. Helena, Calif., C. Mondavi, 1965. 129p. illus. $2.00. LC 67-1978.

A collection of the best from the monthly newsletter *Bottles and Bins* (now published by Krug at the same address). Illustrated by Mallette Dean.

363. Bourbon Institute. **Bourbon Bartender and Chef** (24p.); **Making the Most of Your Bourbon** (24p.); **Book of Bourbon** (28p.).

Collectively, there are over 200 recipes here that cover the addition of bourbon as a flavor enhancer from "soup to nuts." All pamphlets are free, the recipes are exceptionally easy, and beverage mixtures are also indicated.

364. California. Wine Advisory Board. **Adventures in Wine Cookery by California Winemakers**. San Francisco, 1965. 125p. illus.(col.). $2.95pa. LC 65-27205.

These 300 recipes, stressing California wines, were collected from winemakers, from the University of California at Davis, and from Fresno State College. The recipes range from punches to jellies, jams, and desserts.

365. California. Wine Advisory Board. **Easy Recipes of California Winemakers**. San Francisco, 1970. 128p. illus.(col.). $2.50pa. LC 79-134306.

This book is very similar to the preceding one. Each recipe is signed by its contributor. Together with the other five books in this series, these works are among the best buys in the market for cooking with wine and spirits. Each of the books in the series has around 300 recipes that cover the entire menu rather than emphasizing any one particular course.

366. California. Wine Advisory Board. **Epicurean Recipes of California Winemakers**. San Francisco, 1970. 94p. illus.(col.). $3.50pa. LC 79-89800.

These recipes are much more complicated than those in the preceding book, and this difference perhaps accounts for the smaller size and higher price.

367. California. Wine Advisory Board. **Favorite Recipes of California Winemakers**. New York, Essandess Special Editions, 1963. 96p. illus. $1.00pa. LC 63-21635.

First in the series, containing 300 introductory "prize" recipes. A bargain at the price.

368. California. Wine Advisory Board. **Gourmet Wine Cooking the Easy Way**. San Francisco, 1968. 128p. illus. $2.50pa. LC 68-29895.

The only one in the series with unsigned recipes, this book contains tested recipes from producers of convenience foods (e.g., frozen, powdered, freeze-dried, canned, etc.) to which wine was added. Brand names are given at the back. A worthy attempt to put more flavor back into food.

369. California. Wine Advisory Board. **Wine Cookbook of Dinner Menus**. San Francisco, 1971. 128p. illus.(col.). $2.95pa. LC 70-156348.

The emphasis here is on menu preparation and the complementary wine to be served.

370. Chase, Emily. **The Pleasures of Cooking with Wine**. Englewood Cliffs, N.J., Prentice-Hall, 1960. 243p. $4.95. LC 60-13892.

Recipes for all courses, emphasizing California wines. Personal comments accompany the directions. Preparations are relatively easy. Wine charts detail the relationship of wines to foods. The text contains the usual material on selection, storing, and serving wines.

371. Child, Julia, and Simone Beck. **Mastering the Art of French Cooking**. New York, Knopf, 1961, 1970. 2v. illus. $25.00 boxed. LC 61-12313.

The value here is two-fold: information on equipment and products available in America that can be used to reproduce French cuisine, and the detailed description of food preparation, with each step clearly outlined. Emphasis, of course, is not on cooking with alcoholic beverages, but many of the recipes do seem to call for the routine addition of such beverages. Volume Two has a combined index for both books.

372. Comité Interprofessionel du Vin de Champagne. **Cooking with Champagne**. Paris, Lallemand, 1970. 96p. illus.(part col.). free. LC 77-575892.

Translated from the French. Champagne is best with seafood, fish, and desserts, so many of the 117 modern recipes deal with these subjects, although there are also sections on poultry and game. The 26 additional recipes from the past (1712-1833) are from Massalot, Menon, and Carême. A good explanatory chapter gives reasons for using champagne in cooking.

373. **Cooking with Wine**. By the editors of Sunset Books and Sunset Magazine. Menlo Park, Calif., Lane Books, 1972. 80p. illus. $1.95pa. LC 78-180524.

Very similar in format and presentation to the California Wine Advisory Board's series of cookbooks, except that it has fewer pages. The introductory material includes advice on the use of wine in cooking (including a detailed wine cookery chart with suggested quantities of wine to use per serving, ideal for the cook who wishes to improvise) and aids in recipe reading, with names of wines. The 200 recipes, which cover all courses, are geared to California wines. Preparations also include information on making wine vinegar, deglazing, and marinating, as well as special dishes using buffalo or octopus meats. The book concludes with a dozen or so recipes for wine drinks.

374. Fahy, Carole. **Cooking with Beer**. New York, Drake, 1973. 144p. illus.(col.plates). $5.95. LC 72-3266.

Originally published the year before in England, this book stresses the three main beers: lager, ale, and stout. Only the former (lager) is generally available in the United States, so a few of the recipes may be difficult to execute. The 300 recipes cover all courses, including dessert (with a very good date-nut bread). The barbecue section is also recommended. There is historical information, and there are quite a few recipes requiring the mere optional addition of beer for a "special" flavor.

375. Gourmet Magazine. **The Gourmet Cookbook**. Rev. ed. New York, 1965. 2v. illus. $27.50 set.

Here are 5,000 recipes emphasizing glorious eating beyond the regular fare. Cooking with alcoholic beverages is not a main feature; it is just a matter of course with *Gourmet Magazine*.

376. Greenberg, Emanuel. **Gourmet Cooking with Old Crow**. Frankfort, Ky., Old Crow Distillery Co., 1970. 159p. illus.(part col.). free. LC 70-136759.

This is a bargain book, since it is a promotional giveaway. The recipes here are derived from those that used bourbon in the author's 1968 *Whiskey in the Kitchen* (see below). Also included are a few newer items, plus modifications of recipes using other types of spirits.

377. Greenberg, Emanuel, and Madeline Greenberg. **Whiskey in the Kitchen**. Indianapolis, Bobbs-Merrill, 1968. 315p. illus. $10.00. LC 68-29297.

Believing that liquor is an extension of the herb shelf and can thus be used to enhance the flavor of almost everything, the authors present 400 basic recipes for every course. Additional material includes chapters describing each liquor type, what spirits enhance the natural flavors of what foods, tips to brighten a dinner party, the mating of food and spirits, and a history of liquor use since the beginnings of the United States.

378. Hale, William Harlan. **The Horizon Cookbook and Illustrated History of Eating and Drinking Through the Ages**. New York, American Heritage Publ. Co.; distr. by Doubleday, 1968. 768p. illus.(part col.). $16.50. LC 68-15655.

Part One is the historical section, in narrative prose, complete with anecdotes and superb, lavish illustrations. Part Two, the recipe section, includes 600 historical and unusual tested recipes, plus 19 recreated menus from the past. In all, there are 570 illustrations, 110 in color. Most of the recipes are European in origin, and many use alcohol in their preparation. For a history of American eating and drinking, see the companion volume (entry 360).

379. Hallgarten, Elaine, and Dorothy Brown. **Cookery Do**. London, Wines and Spirits Publications, 1972. 162p. illus. $6.00.

Billed as a modern-style recipe book written for the busy housewife, this is an attempt to marry wine with everyday, common ingredients plus certain convenience foods. Washable cover.

380. Hatch, Edward White. **The American Wine Cookbook**. New York, Dover Publications, 1971. 314p. $2.50. LC 76-166428.

This is a reprint of a 1941 edition. At the time of publication, it was pretty daring because it stressed American wines only—and this was shortly after Repeal when the industry was just getting back on its feet. The 700 detailed recipes cover all courses.

381. Hébert, Malcom R. **The First Brandy Cookbook**. San Francisco, Nitty Gritty Productions, 1973. 179p. illus. $3.95pa.

In layout, this is a typical Nitty Gritty book, with colored paper and comic strip art. In addition to 10 pages of menu suggestions and some common recipes, Mr. Hébert stresses inventive new ideas for the creation of recipes. To illustrate this point, there are a large number of recipes that use brandy merely as a flavor enhancer. Some of the better sections deal with appetizers and drinks (cocktails, flips, and punches). New combinations are stressed, but generally these just substitute brandy for some other alcoholic beverage. At the end of the book, there is a brief explanation of brandies. Some of the recipes are complicated, so this is not a book for the cookery novice.

382. Hellman, Renee. **Food and Wine in Europe**. New York, World Publishing, 1972. 128p. $2.25pa. LC 75-187089.

First up in the text is a section on how to budget one's money in Europe. Twenty-two countries are covered (Austria to Yugoslavia), though the biggest section is on France. Arrangement within is: brief description of country; style of eating; types of restaurants; menu terms; national specialties; regional specialties; wine; and a food and drink vocabulary. At the end, there are short vocabularies in six languages for restaurants, food, and drink. Most information was gathered from tourist offices. This is a companion volume to World's "Budget Travel Guides."

383. Ko-operatieve Wynbouwers Vernging Van Zuid-Afrika, Beperkt. **Entertaining with Wines of the Cape**. 4th ed. Paarl, South Africa, 1971. 79p. illus. free. LC 72-184518.

A sharp little book, well illustrated with photographs that describe choosing, cellarage, serving, and cooking with wines from the Cape. Most wines available will be from the Paarl Co-operative. Some recipes are indigenous to South Africa (e.g., Ostrich Eggs).

384. Lewin, Esther. **Stewed to the Gills: Fish and Wine Cookery**. Los Angeles, Nash Publishing, 1971. 165p. illus. $7.95. LC 70-160164.

These easy international recipes (about 225 of them) deal with all types of courses and make use of liquor and beer as well as wine. Arrangement is by type of fish. There are 19 recipes for abstainers (including one using "near beer"). Humor is a little on the flip side.

385. Licensed Beverage Industries, Inc. **Recipe Collector's Roundup: Chance to Choice in Spirited Cooking Techniques**. New York, 1972. 15p. free.

Twenty-six recipes for all courses, using liquor only (no wine or beer).

386. McDonald, Robin. **Cooking with Wine**. London, Allen Lane, Penguin, 1968 (1970, pa.). 144p. illus. $6.00; $1.25pa.

Both British and American measurements are used in the 250 recipes. Beer, cider, and spirits are also included, as well as some preparations for drinks. Emphasis is on French cuisine, with a good section on poultry and savouries. One of the best of the wine cookbooks.

387. Montagné, Prosper. **Larousse Gastronomique: The Encyclopedia of Food, Wine and Cooking**. New York, Crown, 1961. 1101p. illus.(part col.). bibliog. $20.00. LC 61-15788.

Here are 8,500 recipes, including 122 for sole alone. It is also a history of cookery. Alphabetical in arrangememt (by English terms) it has appropriate cross references for English and American terms. Entries for wine may be found by types, country, and grape. Maps are provided. There are diagnoses of wine illnesses in the bottle, and information on how to correct the situation at home. This book is for the skilled cook.

388. Morny, Claude, ed. **A Wine and Food Bedside Book**. Newton Abbot, David and Charles, 1973. 334p. illus. music. £3.00. LC 73-158424.

These are 70 short essays from the late André Simon's *Wine and Food* magazine, itself a ceased publication. It is recommended as bedtime reading, as Harry Yoxall suggests in his foreword, with the book propped on a "happily but lightly filled stomach." The authors include E. M. Forster on sausages, Hilaire Belloc on wine, Osbert Sitwell on stage food, and Cyril Connolly on old restaurants. Several selections by Mr. Simon are also included.

389. Stone, Jennifer. **The Alcoholic Cookbook**. London, Michael Joseph, 1972. 187p. illus. £2.50. LC 72-171480.

A humorous book, with sound recipes, but perhaps to farcical to be taken seriously. What can one do with a "Mother-in-Law's Hot Pot" (cooked in equal quantities of stout and bitter). Imagination gone rampant; best left to the British.

390. Taylor, Greyton H. **Treasury of Wine and Wine Cookery**. New York, Harper and Row, 1963. 278p. illus. $5.95. LC 63-16538.

The author is from the Taylor Wine Company and of Bully Hill fame. Here are 400 recipes plus 150 drink concoctions, along with tips on the service of wine, stemware, and food ideas. There is a superb special section on barbecue cooking, marinades, and canapes (only two of all the cookbooks listed here consider outdoor cooking with alcohol). All recipes were tested at the Taylor Wine Company kitchen. The book is also a work of art, with champagne-tinted paper, sepia ink, and border decorations.

391. Van Zuylen, Guirne. **Eating with Wine**. London, Faber, 1972; distr. Transatlantic Arts. 162p. illus. $8.75. LC 72-193442.

Like many other wine books, this broad survey on a simple level reduces wines to their fundamentals. Otherwise redundant, the book is notable for the inclusion of 24 seasonal menus (recipes given) and likely wines to match. The author, who comes from the Médoc area of France, has also written a good chapter on cheeses: she proposes 50 or 60 cheeses, describing each and detailing the accompanying wines.

392. Wood, Morrison. **Cooking with Wine**. New York, New American Library, 1970. 224p. $1.25pa.

All the recipes here were taken from his three previous books (see below). These are the most outstanding ones, and were originally from his syndicated weekly cooking column "For Men Only." All courses are well covered. This book was originally issued in 1962 as *Specialty Cooking with Wine*.

393. Wood, Morrison. **More Recipes with a Jug of Wine**. New York, Farrar, Straus, and Cudahy, 1956. 400p. $6.95. LC 56-11636.

394. Wood, Morrison. **Through Europe with a Jug of Wine**. New York, Farrar, Straus, 1964. 302p. $7.95. LC 64-21498.

395. Wood, Morrison. **With a Jug of Wine**. New York, Farrar, Straus, 1949. 298p. $6.95.

These three books, spaced more or less eight years apart, detail Wood's wanderings through Europe, the people he met, the sights he saw, the food he ate, and the wine he drank. The most outstanding recipes from these three books were collected into Wood's *Cooking with Wine* (entry 392).

CRAFTS AND COLLECTING

After the wine has been drunk and the food eaten, the wine lover is left with an empty bottle. What to do? Soak off the label, cut up the glass, recork the bottle with different contents, and so forth. This section describes a very small but growing literature on the crafts and collecting possibilities affiliated with alcoholic consumption.

396. Almaden Vineyards. **A Jug and Thou**. San Francisco, 1972. 16p. free.

Suggests imaginative ideas for reusing empty wine bottles and jugs, and explains how to carry out these ideas (e.g., planting a terrarium in a gallon or half-gallon jug, a dinner bell from a burgundy bottle, plus cups and glassware).

397. Anderson, Sonya, and Will Anderson. **Beers, Breweries and Breweriana: An Informal Sketch of United States Beer Packaging and Advertising**. Carmel, N.Y., 1969. 122p. illus. bibliog. price not reported. LC 79-11246.

A catch-all historical work dealing with corporate histories and collecting (trays, bottles, glasses, cans, etc.). The brewing industry in the nineteenth and twentieth centuries was an innovator in advertising. Here are superb photographs and texts for identifying: trays, signs, bottles, pocket knives, cups, reverse glass paintings, matches, letter openers, and so forth.

398. Ash, Douglas. **How to Identify English Drinking Glasses and Decanters, 1680-1830**. London, G. Bell, 1962. 200p. illus. £2.00. LC 62-6020.

399. Ash, Douglas. **How to Identify English Silver Drinking Vessels, 600-1830**. London, G. Bell, 1964. 159p. illus. plates. £3.00. LC 64-9023.

These two works, part of Bell's "How to Identify" series, deal with descriptions of various vessels that were used for drinking. Both are well illustrated and well catalogued, with tradesmen's marks clearly identified, location of deposit, manner of drinking, and other aspects that help to identify the vessels. The section on decanters is particularly interesting, since they are rarely used today except for spirits. The second book is broader in scope, involving even what we today would call soup containers. No duplication in coverage.

400. Dingman, Stanley T. **Wine Cellar and Journal Book**. Richmond, Va., Westover Publ. Co., 1972. 95p. illus. $2.95pa. LC 71-188113.

Tips and advice on starting a wine cellar, plus added pages to use for making journal entries.

401. Escritt, L. B. **The Wine Cellar**. 2nd ed. London, Jenkins; distr. International Publications Service, 1971. 76p. illus. $5.00pa.

A detailed account of how to make and stock a home wine cellar, with advice on serving, tasting, and drinking wine. Other chapters cover a history of casks, bottles, glasses, and wine making.

402. Penzer, Norman M. **The Book of the Wine Label**. London, Howe and Van Thal, 1947. 144p. illus. bibliog. out of print.

Clearly shows the evolution of the wine label, with material on enamels, plates, the classification of designs, lists of names, and ways of identification. Foreword by A. L. Simon. This classic was the first comprehensive account. Whitworth's *Wine Labels* (see entry 404) may be more up to date, and it does have color photographs, but it lacks a bibliography. Penzer furnishes a bibliography, although regrettably it covers only up to 1946.

403. Victoria and Albert Museum. **Bottle Tickets**. London, H.M.S.O., 1958. 32p. illus. (Victoria and Albert Museum Publication, No. 44). £0.25.

"Bottle tickets" was the first name that wine labels went under, for they were simply items attached to bottles to identify them. Twenty-six pages of this catalog are photographs. For more information, write to the Wine Label Circle (Britain).

404. Whitworth, E. W. **Wine Labels**. London, Cassell, 1966. 63p. illus.(part col.). (Collectors' Pieces, 8). £0.75. LC 66-74790.

This presents the historical development of labels (*not* the paper kind). The reader is told where he may view the rarer pieces, and an identification list is given. Most of these "decanter labels" are silver. Whitworth shows how to date labels, describes types (e.g., shield, vine and tendril, goblet, etc.), and lists up to 20 makers. Material also includes mother-of-pearl, Sheffield plate, and enamel (as does Penzer, entry 402). Bin labels, which are used to identify wines stored in wine cellars, are also discussed. They are not as elaborate as wine labels.

HEALTH ASPECTS

To mitigate the view of alcohol as merely Epicurean or Lucullan pleasure, this section deals with the therapeutic health aspects of alcoholic beverages. More information can be obtained from the Society of Medical Friends of Wine. (See p.162.)

405. Lucia, Salvatore, ed. **Alcohol and Civilization**. New York, McGraw-Hill, 1963. 416p. illus. $3.95pa. LC 63-18706.

Dr. Lucia is a one-man organization dedicated to the therapeutic nature of alcohol. In this symposium (organized by the School of Medicine, University of California), 21 authors examine alcohol as a drug, a food, and a chemical. Nearly all are medical doctors, so the dominating sections deal with the effects of alcohol on the mind and on the body, along with social and legal implications, and intoxication.

406. Lucia, Salvatore. **A History of Wine as Therapy**. Philadelphia, Lippincott, 1963. 234p. illus. $6.50. LC 63-14618.

A straightforward account of those times when wine was an assistance to mankind.

407. Lucia, Salvatore. **Wine and Your Well Being**. New York, Popular Library, 1971. 160p. illus. bibliog. $0.95pa. LC 70-143901.

Medical applications of wine. This book, very similar to the author's *Wine as Food and Medicine*, in effect updates it (see entry 409).

408. Lucia, Salvatore, ed. **Wine and Health**. Menlo Park, Pacific Coast Publishers, 1969. 85p. illus. $5.95. LC 78-90896.

Proceedings of the First International Symposium on Wine and Health, held in Chicago in 1968. Papers deal with emotional tension, diabetic use, wine in hospitals and nursing homes, wine pigments, heart disease, and cancer.

409. Lucia, Salvatore. **Wine as Food and Medicine**. New York, Blakiston, 1954. 149p. bibliog. $3.00. LC 54-7397.

Lucia concentrates on the chemistry and physiological effects of wine (respiration, cardiovascular, kidneys, diabetes, neuromuscular), plus its medical use for digestive and nutritional benefits (strong in minerals). Largely updated by his *Wine and Your Well Being* (see entry 407).

410. Shackleton, Basil. **The Grape Cure: A Living Testament**. London, Thorsons, 1969. 128p. £1.00. LC 76-450654.

This is one man's account of how he recovered from illness by the use of wine. It is based on his *About the Grape Cure* (London, Thorsons, 1962), a book half the length of the present volume; the 1969 work reflects additions made during the seven-year period between the books.

411. Van Essen, William. **A Man May Drink**. By Richard Serjeant (pseud.). London, Putnam, 1964; distr. Queenswood. 191p. illus. bibliog. $3.95. LC 65-9608.

Particularly good on the physiological effects. This is but one aspect of the pleasures that derive from drinking, but Van Essen makes it a big point.

HISTORY AND HISTORICAL WORKS

The books in this section have been carefully selected out of the hundreds available (as either social or economic histories). Temperance, Prohibition, and Repeal together account for many such titles. Since these particular books are mainly political or economic in tone, we have listed only a few important studies from that genre. Further investigation into these fields will be found by consulting the books listed here. Typical material includes Henry W. Lee's *How Dry We Were* (Prentice-Hall, 1963). Histories of frivolous drinking, along more or less humorous lines, have also been excluded, such as Douglas Sutherland's *Raise Your Glasses: A Lighthearted History of Drinking* (London, Macdonald, 1969), as well as specific, scholarly works of a very detailed nature, such as Om Prakash's *Food and Drinks in Ancient India, from Earliest Times to 1200 A.D.* (Delhi, Munshi Ram Manohar Lal, 1961). Regional histories that are part of a larger general study will be found under the appropriate geographic area above, as will be found corporate histories or histories of a particular breed of wine. Works on pubs and inns are included in a separate part of this chapter.

412. Adlum, John. **A Memoir on the Cultivation of the Vine in America, and the Best Mode of Making Wine**. Washington, D.C., 1823; repr. Hopewell, N.J., Booknoll Reprints, 1971. 144p. $12.50.

The first book printed in the United States on wine, with the addition of a biography and bibliography by Dorothy Manks, from "Huntia II"; a facsimile of Thomas Jefferson's letter of April 1823 to Adlum regarding his book and some of his wine; and Adlum's "Petition to Congress," April 30, 1828, regarding the book and his pioneering efforts in making wine from native grapes.

413. Allen, H. Warner. **A Contemplation of Wine**. London, Michael Joseph, 1951. 232p. £2.00.

As with most of Allen's writings, this is a potpourri of thoughts concerning mini-histories of cocktails, Roman wines, corkscrews, and other accoutrements of alcoholic beverages. He also presents sketches of the noted chef Brillat-Savarin and of two of Allen's contemporaries, George Saintsbury and André Simon.

414. Allen, H. Warner. **A History of Wine: Great Vintage Wines from the Homeric Age to the Present Day**. New York, Horizon Press, 1962. 304p. illus. bibliog. $10.00. LC 62-9746.

Allen describes in great detail the rediscovery of the Greek airtight amphorae used to preserve wine against spoilage. The long series of experiments culminated with vintage port in the latter half of the eighteenth century. The glass bottle replaced the earthenware jar. This scholarly but easy-to-read book covers mainly up to 1900, with a few concluding pages to 1960.

415. Asbury, Herbert. **The Great Illusion: An Informal History of Prohibition**. Garden City, N.Y., Doubleday, 1950. 344p. bibliog. LC 50-10358. out of print.

A popularly written history that perhaps spends too much time on the evils that occurred, with a long concluding chapter about the end of Prohibition. Most of the histories of this period (and this is as good an example as any) pay no attention to Prohibition's subsequent effects on the treatment of alcoholic beverages by government legislation and regulation (e.g., local option, state monopolies, etc.).

416. Beveridge, N. E. **Cups of Valor**. Harrisburg, Pa., Stackpole Books, 1968. 106p. illus. $6.95. LC 68-29595.

This slender effort concerns the obtaining of drinking alcohol by America's armed forces over the years since the Revolution. It is colloquial in style, in describing the liquor lore of the Army and Navy. Anecdotes relate alcohol's use in easing the pain of battles and fatigue, plus regimental celebrations. Interesting concoctions were made as the result of a lack of proper ingredients (such as the chicken marengo of Napoleon's chef). There are 33 contributed recipes from the eras of the Revolution, the Civil War, the Indian wars, the Spanish-American War, and the two World Wars. Index to recipes and ingredients only.

417. Bishop, George V. **The Booze Reader: A Soggy Saga of Man in His Cups**. Los Angeles, Sherbourne Press, 1965. 288p. illus. $4.50. LC 65-26329.

This is an eclectic dossier of drink and drinking. Topics covered (historically) include American whiskey, Scotch, why people drink, the A.A., tipping the waiter, beer, advertising, sex, wines, drunks, disease, local option, bars, and cocktails. Much of it deals with beers and spirits. This is undeniably a hodge-podge collection. Over 30 historical cartoons and paintings are reproduced on the 16 plates.

418. Carr, Jess. **The Second Oldest Profession: An Informal History of Moonshining in America**. Englewood Cliffs, N.J., Prentice-Hall, 1972. 250p. illus. index. bibliog. $7.95. LC 70-172273.

A totally fascinating book about the illicit distillation of whiskey. The author, a native of mountainous western Virginia, has produced a complete study which traces the history of moonshining from colonial times through the 1960s. His richly detailed chapters on producing "mountain dew" and on the types of stills in use are completely absorbing. One can't go wrong by following Chapter Seven, "Moonshining Methods and Equipment," or Chapter Twelve, "Types of Stills and Their Application." However, if one reads the other chapters one will find what happens to someone who "runs off a little brew now and then."

419. Carosso, Vincent. **The California Wine Industry, 1830-1895**. Berkeley, University of California Press, 1951. 241p. bibliog. LC 51-62531.

A scholarly study of the early formative years, emphasizing hardships.

420. Chidsey, Donald Barr. **On and Off the Wagon: A Sober Analysis of the Temperance Movement from the Pilgrims through Prohibition**. New York, Cowles, 1969. 149p. bibliog. $6.95. LC 69-19997.

This is the story of the corner saloon (bar, tavern, taproom, nightclub, etc.) and of people's attitudes toward private and public drinking. Material covered also includes the W.C.T.U., the Whiskey Rebellion, and the Anti-Saloon League. This is a popularly written account by a popular writer (who has authored over 50 books).

421. Francis, A. D. **The Wine Trade**. London, Adam and Charles Black, 1972. 353p. maps. index. bibliog. £4.24. LC 73-158015.

The wine trade is a worldwide phenomenon, involving large transactions from North Africa, Europe, and Australia. This is a British economic history text, more scholarly than oenological. Because coverage is from Roman Britain to the end of the nineteenth century, it does not update Simon's monumental works, and indeed duplicates much information. This is one of the "Merchant Adventurers" series, which also offers books on timber and whaling.

422. Grace, Virginia R. **Amphoras and the Ancient Wine Trade**. Princeton, N.J., American School of Classical Studies at Athens, 1961. [32] 69p. illus. $0.50pa.

As one of a series dealing with the excavations of the Athenian agora, this small book consists mainly of pictures of wine containers, with little text.

423. Gray, James H. **Booze: The Impact of Whisky on the Prairie West**. Toronto, Macmillan, 1972. 243p. illus. $7.95. LC 73-159254.

The author traces the evolution of the Temperance-Prohibition movement in Canada. He describes the big business that developed with "rum runners." Most of the action takes place in Gray's beloved Prairie Provinces, which were legally dry from 1914 to 1924. The government of Canada made Repeal contingent upon the establishment of the provincial Liquor Control Commissions.

424. Haraszthy, Agostin. **Grape Culture, Wines and Wine Making, with Notes upon Agriculture and Horticulture**. New York, Harper, 1862; repr. Hopewell, N.J., Booknoll Reprints, 1971. 420p. illus. $20.00. LC 73-161074.

This is the first important California winebook. Haraszthy, called "the father of modern California viticulture," went to Europe for cuttings of fine grapes, and brought back over 100,000 cuttings of 300 varieties, which formed the base of California's modern industry. This book is his story of scientific findings and applications of European stock to California conditions.

425. Henderson, Alexander L. **The History of Ancient and Modern Wines**. London, Baldwin, Craddock and Joy, 1824. 468p. tables. LC 8-21692.

Only the first 125 pages are devoted to the ancient world (management of vintage, amphorae, types of wines and banquets). Modern wines (i.e., from around the year 1800) are covered by surveys of existing conditions in European countries. The keeping and storage of wines are discussed, along with adulteration and medical properties. Statistical material and economic information are scattered throughout. The appendix deals with fermentation, the evaporation rate of alcohol, the distribution of sales and vintages since 1780, and some early calculations about import-export matters. Unfortunately, there are no illustrations.

426. Hyams, Edward S. **Dionysus: A Social History of the Wine Vine**. New York, Macmillan, 1965. 381p. illus. maps. $10.00. LC 65-15581.

An exemplary history of viticulture, with 130 reproductions of paintings and other art works, maps, eight color plates, and a good index. As a basic historical work concerning the consumption of wine, the book details *vitis vinifera*: how it was cultivated in the past, what sort of wines were drunk, how they were made and taxed, how all of this affected peoples' lives. This account of a single plant species through 80 centuries is unique in literature, yet the author's style is cloying and his documentation is sketchy. The appendix deals with *Vitaceae*.

427. James, Margery K. **Studies in the Mediaeval Wine Trade**. Oxford, Clarendon Press, 1971. 232p. illus. $13.00. LC 75-860197.

A collection of scholarly essays, edited by Elspeth M. Veale, with a perfunctory introduction by E. M. Carus-Wilson. Topics covered are mainly the economic influences, the political scene between England and France (most notably the possession of Bordeaux), and the trade between England and Portugal.

428. Merz, Charles. **The Dry Decade**. Seattle, University of Washington Press, 1969 (c.1931). 343p. index. bibliog. (Americana Library Series, 13). $9.50; $3.45pa. LC 71-8894.

Originally published by Doubleday in 1931, this reprint can be criticized for being too contemporary, since it was written while the decade was closing. Beginning with the history of the Prohibitionists in the United States before the 18th Amendment, Merz examines the political causes and forces that created Prohibition. Looked at are: the formula used, the enforcement, the propaganda, the criminals, the opposition, the Hoover program, and various appeals from the states. His concluding chapter deals with "the position in 1930," at which time the question of Repeal was not resolved. Additional matter in the appendices covers the adoption of Prohibition laws, the Senate and House votes on the 18th Amendment, the ratification vote by each state, the text of the National Prohibition Act (Volstead Act), the amounts appropriated by Congress for enforcement, the numbers of arrests, seizures, and criminal prosecutions, and the results of state referenda on prohibition questions from 1920 to 1930.

429. Ordish, George. **The Great Wine Blight**. New York, Scribners, 1972. 237p. illus. bibliog. $7.95. LC 72-537.

A fascinating account of the Phylloxera blight on European vineyards, and the subsequent attempts to eradicate the pest. Chapters describe the first bumbling, panic-stricken efforts to find a remedy (e.g., hoeing). Every conceivable cure was tried, including human urine. Grafting was delayed 15 years because of opposition. In France alone, 11 *billion* cuttings were needed to "reconstitute" the vineyards.

430. Peninou, Ernest P., and Sidney Greenleaf. **A Directory of California Wine Growers and Wine Makers in 1860**. Berkeley, Calif., Tamalais Press, 1967. 84p. illus. index. bibliog. $15.00. LC 67-8058. (Edition of 450 copies).

Much biographical and historical detail is given "to commemorate the names of those Californians who more than a century ago were engaged in the growing of grapes and the making of wine." It lists 262 persons, and most of the data come from the Census of 1860. Information concerns acreage, gallons of wine on hand, pounds of grapes sold, and money received for grapes. Sixty names were arrived at from a variety of sources (county histories, agricultural societies' transactions, etc.). Only six sites that were occupied in 1860 are still producing wine. Arrangement is by county, then alphabetical by grower. There are four black and white illustrations of wine labels from the period.

431. Seltman, Charles. **Wine in the Ancient World**. London, Routledge and Kegan Paul, 1957. 196p. illus. bibliog. $5.95. LC 58-16228.

The author describes wine-drinking habits of the ancients, with emphasis on Greece and Rome. He guesses at what ancient wine tasted like, and provides a description of various illustrated amphorae. Other material includes mention of the gods and saints of wine, the history of the grape in the Old and New Testaments, the course of wine in the Homeric epics, a description of a fifth century B.C. Athenian banquet as contrasted to a vulgar Roman feast, and some detail on the Italian, French, and German beginnings of wine after the collapse of the Roman Empire.

432. Simon, André L. **Bottlescrew Days: Wine Drinking in England During the 18th Century**. New York, Finch Press, 1971. 273p. illus. index. $14.00.

This is a reprint of the 1926 original edition published by Duckworth in England. Simon was dedicated to the belief that a greater knowledge of wine would result in a greater appreciation of its virtues. From the days of Queen Anne, he investigates and tells the story of smuggling, excise taxes, customs duties, the shipping of French wines during those troubled times, plus details on port, the wine of Spain, madeira, the Canaries, Italy, and Germany. A description of wine glasses, wine labels, and bottlescrews follows, along with a concluding chapter on drinking songs and toasts.

433. Simon, André L. **The History of the Wine Trade in England**. London, Holland House, 1964. 3v. bibliog. $30.00.

This is a reprint of the 1906 edition, plus a new index. A good, scholarly work, leaning toward the economic history side.

434. Simon, André L. **Wines and Spirits: The Connoisseur's Textbook**. New York, Finch Press, 1961. 194p. illus. $14.00. LC 61-19689.

A reprint of the 1919 edition, which was originally a text for the trade. All encyclopaedic matter is covered here, and there are chapters for port, sherry, wines, madeira, marsala, liqueurs, beer, cider, water (!), and California wines (15p.). Each section is a mini-history of development up to around 1919. Supplementary material includes the care of wine (buying, keeping, decanting, serving) and a list of port shippers and vintages.

435. Thudichum, John, and August Dupré. **A Treatise on the Origin, Nature and Varieties of Wine: Being a Complete Manual of Viticulture and Oenology**. New York, Macmillan, 1872. 760p. illus. LC 8-21695. out of print.

Written in the period immediately following the 1867 invasion of the Phylloxera louse, this text deals at length with the indigenous vines of Europe in relation to history, soil, general principles, climate, medical qualities, vintage and vinification. Extensive chapters on the chemistry of alcohol detail the composition of ethers, acids, sugars, and other components. All the vines in Europe are covered exhaustively in the next 500 pages, and attempts are made to classify wine types. This is a basic classic that deserves reprinting.

436. Younger, William A. **Gods, Men and Wine**. Cleveland, World Publ. Co., 1966. 516p. illus. bibliog. $15.00. LC 66-17305.

This is probably *the* definitive one-volume history of wine. When Younger died in 1961, the book was completed but not corrected. Most of the history is Greek and Roman, since viticulture then staggered until the eighteenth century. He called on a wide range of source material, and some of this has been incorporated into the eight glossary appendices: wines of Egypt and the ancient Middle East; ancient Greek wines; ancient Roman wines; mediaeval wines; mediaeval English vineyards; Renaissance wines; and tables of measures and money for both the mediaeval and the ancient worlds. In addition, Younger speculates on the beginnings of winemaking. John N. Hutchison contributed an 11-page chapter on wines in America. Both the illustrations (color plates) and the bibliography are superb.

LITERARY AND MUSICAL REFERENCES

Alcoholic beverages have played a major role in the awakening of the senses to an appreciation of the fine arts and to the heady feeling of omnipotence.

437. Fadiman, Clifton, ed. **Dionysus: A Case of Vintage Tales about Wine**. New York, McGraw-Hill, 1962. 309p. illus. $6.50. LC 62-19244.

A collection of stories, anecdotes, and facetiae about wine, all of it drawn from the literature of the world.

438. Gordon, Alvin J. **Of Vines and Missions**. Flagstaff, Ariz., Northland Press, 1971. 87p. illus.(col.). $9.70. LC 71-150686.

A very posh book of poems celebrating wine and grapes. Exceptionally well illustrated.

439. Healy, James N., comp. **Ballads from the Pubs of Ireland**. 3rd ed. Cork, Mercier Press, 1968. 142p. illus. £0.65pa. LC 73-481.

A collection of sheet music songs meant to be sung while imbibing.

440. Hogg, Anthony, comp. **Wine Mine**. London, Souvenir, 1970. 233p. illus. £2.10. LC 77-31905.

A hodge-podge collection of wine facts and wine lore, with copious pictures and songs.

441. Scott, James Maurice. **The Man Who Made Wine**. New York, Dutton, 1954 (c.1953). 124p. illus. LC 54-10315 rev. out of print.

This is a novel—probably the only one in English literature thus far—that deals with an account of winemaking and life in the vineyards of Bordeaux. Well researched.

442. Shay, Frank. **My Pious Friends and Drunken Companions, and More Pious Friends and Drunken Companions**. New York, Dover, 1961. 235p. illus. $6.50pa. LC 61-66216.

This is a reprint of two books dealing with the drinking songs most applicable to students.

443. Taylor, Sidney B., comp. **Wine . . . Wisdom . . . and Whimsy**. Portland, Ore., Wine Press Publ. Co., 1969. 160p. illus.(part col.). $7.95. LC 70-98178.

Another compilation of anecdotes and fiction, with gorgeous illustrations. Well worth the price despite the slimness of the volume.

PERSONAL NOTES OF CONTEMPLATION

Whether it stems from a finer appreciation of wines and the good life or from a quieter existence in Britain than in America, British writers have produced more and better writings on the contemplation of alcoholic beverages than any other national group. Full of anecdotes, reflections, tastings (see also the section on tastings), facetiae, and so forth, these books, while not technical, do presuppose a knowledge of alcoholic beverages. Indeed, the books often give the impression of being letters from one friend to another; hence, these books are usually not recommended for the novice.

With the rise in prices and the disappearance of the literate wine merchant (because of the dictates of hard-nosed economics and business within the industry), the "British Baroque School of Wine Writings" virtually ended by the late 1960s. The first such book was Saintsbury's, which advocated the lessening of spirits with meals, the substitution of fine wines, and the establishment of a status for wine appreciation on its own—just to drink wine and to savor it. Other such writings, in addition to what is to be found below, are long out of print but still worth reading. Many deal with the great vintages of pre-Phylloxera wines of the past, and, of course, the great vintages of the 1920s.

A quick listing of these older works would include the books of H. Warner Allen (*Natural Red Wines*, 1951; *Through the Wine Glass*, 1954), C. W. Berry (*In Search of Wine*, 1935), and Ian M. Campbell (*Wayward Tendrils of the Vine*, 1948; *Reminiscences of a Vintner*, 1950).

444. Allen, H. Warner. **The Romance of Wine**. New York, Dover, 1971. 264p. illus. $2.50pa. LC 71-166424.

This is a reprint of the original 1931 edition. These are tales of wine drinking, appraisals, and tastings by a British expert of the old school.

445. Allen, H. Warner. **Through the Wine Glass**. London, Michael Joseph; distr. Transatlantic Arts, 1954. 244p. $3.75. LC A56-5653.

Very much a continuation of *The Romance of Wine* and the more historical *A Contemplation of Wine*.

446. Gould, Francis Lewis. **My Life with Wine**. St. Helena, Calif., Krug Winery, 1972. $4.95.

As a writer and lecturer on wine for 40 years, the 90-year-old Mr. Gould can give authoritative advice on a lot of matters. This autobiographical work is resplendent with such phrases as "to serve a Romanee Conti with a hot dog or hamburger is not the proper caper." He knew and drank noble vintages with the socially elite, such as personal friends André Simon and Isadora Duncan. A nostalgic and humorous book from the writer of the newsletter *Bottles and Bins* (since 1949).

447. Healy, Maurice. **Stay Me with Flagons**. London, Michael Joseph, 1940, 1963. 262p. £1.15.

This is a diverting, idiosyncratic book that is a frank assessment of classical wines by a man who was then a gifted amateur. It is, however, only incidentally informative.

448. Holland, Vyvyan Beresford. **Drink and Be Merry**. London, Victor Gollancz, 1967. 173p. maps. tables. £1.25. LC 67-77010.

The author, once vice-president of the Circle of Wine Writers, has collected a number of personal recollections that form a knowledgeable book. The bits and pieces of information are gathered in colloquial and anecdotal style. Unfortunately, the commonsense wit tends to go off on a tangent from time to time (Chinese wines, vintage olive oil, vinegar, and even water). There is a pre-occupation with sardines.

449. Marcus, Irving H. **Lines about Wines**. Berkeley, Calif., Wine Publications, 1971. 214p. illus. $5.75.

A collection of almost 100 editorial pieces written for *Wines and Vines* (Marcus was its former editor). Material ranges from cellar practices to public attitudes toward wines. These informative, chatty columns were written from 1956 on, primarily for California winemakers.

450. Mendelsohn, Oscar A. **Drinking with Pepys**. New York, St. Martin's, 1963. 125p. illus. ports. $4.75. LC 64-2715.

Mendelsohn is both a chemist and a writer. Here he has collated all of Pepys' allusions to alcohol, to possets and syllabubs, cellarage, and cooperage. Through Pepys, then, he covers taverns, tap-houses, the cellars of Restoration London, and some vineyards of Europe and of historical antiquity. On his own, he writes a description of the chemistry of fermentation and distillation.

451. Pellegrini, Angelo M. **Wine and the Good Life**. New York, Knopf, 1965. 306p. $5.95. LC 64-19105.

Taking the position that he is an out-and-out sinner and scoundrel, Mr. Pellegrini tries to justify his existence and that of others by declaiming the joys of living well. These essays delve into plenty of food and wine, and evolve into a way of life where one lives only for wine, cheese, bread, and love.

452. Ray, Cyril. **Compleat Imbiber**. New York, Taplinger, 1963. 208p. illus. $8.95. LC 62-13656.

This book is a collection from the first few years of the periodical *Compleat Imbiber*. It contains many miscellaneous poems devoted to the subject of wine appreciation, as well as drawings, recipes, and various anecdotes. Both major and minor contributors to wine writing are well-represented.

453. Ray, Cyril. **In a Glass Lightly**. South Brunswick, N.J., A. S. Barnes, 1969 (c.1967). 193p. illus. $4.95. LC 75-83493.

Ray examines all the components of a glass of wine, in a highly personal style. Good illustrations by Quentin Blake.

454. Saintsbury, George. **Notes on a Cellar Book**. New York, St. Martin's, 1964. 231p. $4.25. LC 64-2699.

Originally published in 1920 by an authority on the art of good living, this is a classic series of reminiscences based on the premise that wine is what "God sends to make men glad." Literary allusions abound in this book, which covers wine, beer, spirits, and cider. Bottles and glasses are discussed, as well as starting a cellar and classic menus. His erudite style led the way for today's modern wine appraisal.

455. Saucier, Ted. **Bottoms Up**. New and rev. ed. New York, Greystone Press, 1962. 288p. illus. $12.95. LC 62-12955.

A general book on the consumption of alcoholic beverages, with superb illustrations by 11 American artists (including Arthur William Brown). Even the cover is provided with detailed illustrations.

456. Simon, André L. **The Art of Good Living**. 2nd ed. London, Michael Joseph, 1951 (c.1929). 197p. illus. out of print.

One of the many readable books from this prolific author. His philosophy is expounded here in great detail: "Thought and care in the matter of eating and drinking offer far greater rewards than mere satisfaction of appetite." He would wish to eliminate the snobs and the restaurants. Much material also covers food generally, and there are glossaries of definitions.

457. Simon, André L. **In the Twilight**. London, Michael Joseph, 1969. 182p. £1.75. LC 71-447622.

André Simon died in 1970, after 93 years of life. This book serves as his epitaph: it describes his life in wine, reflections, and second thoughts on what he had said over the past 70 years. A masterpiece.

458. Waugh, Alec. **In Praise of Wine and Certain Noble Spirits**. New York, William Morrow, 1959. 304p. $2.95pa. LC 59-11712.

A very readable autobiography by a well-travelled novelist; it is essentially a paean to wine. He begins with his own discovery of wine, then provides a history of winemaking and a discussion of individual wines found on his travels—e.g., of a $1.30 New York State port, he writes, "The first sip was one of the biggest shocks my palate has sustained . . . it bore no resemblance to anything" (p. 101). He concludes with a serious discussion of fallacies about wine. Tasting notes, food notes, and menus are scattered throughout.

PUBS AND INNS

Anyone the least bit interested in reading about alcoholic beverages would probably want to extend his horizons to the places where consumption is the highest: the country pub or inn.

Older works, which are out of print but which could be called "classics" and are therefore subject to future reprint, include: George Ade, *The Old Time Saloon* (New York, Ray Long and Richard Smith, Inc., 1931); E. Field, *The Colonial Tavern* (Providence, R.I., Preston and Rounds, 1897); V. Efrom, "The Tavern and the Saloon in Old Russia," in R. McCarthy, ed., *Drinking and Intoxication* (New York, Free Press, 1959); F. W. Hackwood,

Inns, Ales, and Drinking Customs of Old England (London, T. Fisher Unwin, 1909); and Mass Observation, *The Pub and the People* (London, Victor Gollancz, 1943).

The books listed below share common characteristics; hence, the annotations have been kept short to avoid redundancy. There is little duplication (some have maps and plans); and the treatment is historical and regional. Titles are self-explanatory, and this is a highly selective list of in-print materials. For an interesting work on current bars and behavior, see entry 81.

459. Crawford, Mary Caroline. **Among Old New England Inns**. Detroit, Singing Tree Press, 1970. 381p. illus. (Little Pilgrimages Series). $14.00. LC 76-107629.

An account of little journeys and pilgrimages to various quaint inns and pubs of colonial New England. This is a facsimile reprint of the 1907 edition.

460. Dallas, Sandra. **No More Than Five in a Bed**. Norman, University of Oklahoma Press, 1967. 208p. illus. map. $5.95. LC 67-15587.

An illustrated account of Colorado hotels during frontier days.

461. Earle, Alice Moore. **Stage Coach and Tavern Days**. New York, B. Blom, 1969. 449p. illus. $5.75. LC 70-81558. Also published by Haskell House, 1968. LC 68-26351.

The importance of stage coaches should not be underestimated. They were the only available public transit for over 100 years. The tavern became important because it was the way-station on a long and arduous journey, a place that provided food and accommodation, news and information. This reprint of the 1900 edition restores to availability an important classic that should be read by all persons interested in social history. Its significance lies in the fact that it was written toward the end of the stage-tavern era, when sources, illustrations, and personal knowledge were still available and freshly recalled. Coverage is about equally split between coach details and the hostel. There are many architectural and skeletal drawings.

462. Endell, Fritz A. G. **Old Tavern Signs**. Detroit, Singing Tree Press, 1968. 303p. illus. bibliog. $11.50. LC 68-26572.

An illustrated history of inns and innkeeping, originally published in 1916. This is a facsimile edition.

463. Farnham, C. Evangeline. **American Traveller in Spain**. New York, AMS Press, 1966. 58p. bibliog. notes. $10.00. LC 77-168007.

This was originally part of a Ph.D. dissertation at Columbia University (1921); it was subsequently published as Volume 24 in Columbia's Studies in Romance Philosophy and Literature. Spanish inns of the period from 1776 to 1867 are covered. Unfortunately, there are no illustrations, and the price is rather steep.

464. Firebaugh, W. C. **Inns of Greece and Rome**. New York, B. Blom, 1972. 271p. illus. $15.00. LC 76-175878.

This excellent history, which really extends "from the dawn of time to the Middle Ages," is a reprint of the 1928 Pascal Covici edition. The social life and customs of Greeks and Romans are also examined.

465. Firebaugh, W. C. **Inns of the Middle Ages**. New York, B. Franklin, 1971. 274p. illus. $15.00. LC 24-22961.

Written before Firebaugh's book on Greece and Rome (this is a reprint of the 1924 Pascal Covici edition), this history details only the period of the Middle Ages.

466. Freeland, John Maxwell. **The Australian Pub**. Melbourne, Melbourne University Press; distr. by Cambridge University Press, 1966. 229p. illus. tables. bibliog. $9.20. LC 66-25977.

A definitive history, actually a model of its kind, with superb illustrations and tables detailing consumption and guests.

467. Guillet, Edwin C. **Pioneer Inns and Taverns**. Toronto, Ontario Publ. Co.; distr. University of Toronto Press, 1964. 5v. in 2. illus.(part col.). maps. $50.00 boxed.

First published in five volumes from 1954 to 1962, these epic tomes cover the main pioneer routes in Ontario, Quebec, and New York State. Detailed reference is made to metropolitan Toronto and Yonge Street to Penetanguishene, and to the New York-Buffalo route via the Hudson River and the Erie Canal. There is a concluding estimate of the position of the innkeeper in community life, plus a large section (the entire original fifth volume) on the origins of tavern names and signs in Great Britain and America.

468. Haas, Irwin. **America's Historic Inns and Taverns**. New York, Arco, 1972. 182p. illus. $8.95. LC 66-22320.

A good, general history, but it attempts to cover too much in one volume. Useful as an introduction.

469. Hartwig, Wolfgang, and Winfried Löschburg. **Great Inns of Europe**. New York, Hart, 1972. $6.95.

A translation from the German detailing the majestic stature of the fine eating and sleeping places currently in operation in Europe.

470. Hartwig, Wolfgang, and Winfried Löschburg. **Historic Inns of Europe**. New York, Hart, 1972. 135p. illus.(part col.). $6.95.

A translation from the German which discusses the historic inns of Europe, some of which are still functioning.

471. Jones, Vincent. **East Anglian Pubs**. New York, Hastings, 1965. 152p. illus. $3.50. LC 65-8987.

One of many books detailing the pubs of different regions of old England, published originally by Batsford in London. Good illustrations by Leo Gibbons-Smith. See also entries 472, 477, and 478.

472. Keeble, Richard. **Surrey Pubs**. New York, Hastings, 1965. 148p. illus. $3.50. LC 65-9009.

One of the regional surveys of pubs in England. Good illustrations by Joan Charleton. See also entries 471, 477, and 478.

473. Lathrop, Elise L. **Early American Inns and Taverns**. New York, Blom, 1968. 365p. illus. bibliog. $12.50. LC 68-20234.

This is a reprint of the 1926 Tudor edition, and naturally the bibliography only goes up to that time. Arrangement is by the New England states. Of historical interest, since this survey concerned only the inns still in existence in 1926.

474. Licensed Beverage Industries. **The American Tavern: Hospitality and Historic Tradition**. New York, 1973. 24p. illus. bibliog. free.

A short historical account of pubs in America, dealing mainly with those in New England, the West, and New York City, and covering laws, rules, owners, and use as meeting places. It is crammed with facts about "minding your P's and Q's," the naming of towns after pubs, the origin of "tip," "yards" of ale, and so forth.

475. Richards, Timothy M., and James Stevens Curl. **City of London Pubs: A Practical and Historical Guide**. New York, Drake, 1973. 216p. illus. maps. bibliog. $6.95. LC 72-11267.

In Britain, the pub was the hub of social activities. The authors, with anecdotes and enthusiasm, follow the history of 1,153 taverns in the seventeenth century City of London (all within the one square mile) down to the present day, when 162 interesting taverns still exist. This is a complete guide to each of these extant establishments, arranged into 10 walking tours. Maps and 30 photographs make this a good book for the traveller.

476. Richardson, Sir Albert E., and H. D. Eberlein. **The English Inn, Past and Present**. New York, Blom, 1968. 307p. illus. maps. $12.50. LC 68-56499.

This reprint of the 1925 classic restores the book to the world of print. This is a review in an historical and social context (mainly the latter), with detailed maps and architectural plans.

477. Tubbs, Douglas B. **Kent Pubs**. New York, Hastings House, 1966. 144p. illus. $3.50. LC 66-74219.

Another in the series of regional histories, with excellent drawings by Alan F. Turner. See also entries 471, 472, and 478.

478. Walkerley, Rodney Lewis. **Sussex Pubs**. New York, Hastings House, 1966. 175p. illus. $3.50. LC 66-70771.

Fourth in this special series. See also entries 471, 472, and 477.

479. White, Arthur S. **Palaces of the People**. New York, Taplinger, 1970 (c.1968). 180p. illus. $5.50. LC 70-107011.

The broad sprawling social history of "commercial" hospitality is less interesting than the other books in this series, but it is an enjoyable introductory text.

480. Yoder, Paton. **Taverns and Travelers: Inns of the Early Midwest**. Bloomington, Indiana University Press, 1969. 246p. illus. bibliog. $6.95. LC 70-85104.

A most scholarly book with facsimile reproductions and a good bibliography. Probably the only such book to cover the American Midwest, it serves as an excellent "sightseeing" guide.

TASTING AND EVALUATION GUIDES

The most important aspect of alcoholic beverages is their consumption. Yet there are very few guides to tasting, and perhaps only a bit more in the way of consumer education. How does one "taste" beer, or wine, or whiskey? Notes on these subjects are scattered widely in the literature, and the reader here is referred to the "Personal Notes of Contemplation" section as the one most likely to contain additional tasting and evaluation notes.

481. Amerine, Maynard A., Rose Marie Pangborn, and Edward B. Roessler. **Principles of Sensory Evaluation of Food**. New York, Academic Press, 1965. 602p. illus. $15.00. LC 65-22766.

A highly technical but readable book. Sensory analysis is arrived at through knowledge of sensory physiology, psychology, and perception. Outside aid is provided for by careful statistical analysis of the data. The sole difficulty appears to be in measuring consumer acceptance. The complexity of a wine bouquet and taste are examined.

482. Baker, John V., ed. **The Paragon of Wines and Spirits**. Vol. 1. London, privately printed by the editor, 1972. various paging. illus. $7.25.

Discusses in some detail over 70 wines that won awards at the 1971 international competition (first held in 1966) conducted by the British Club Oenologique. All entries had to be subjected to the Massel Quality Index test and a score of 100 (out of 130) is the necessary minimum before a wine could be considered for an award. On the M.Q.I., the tasting notes account for a maximum of 54 points, while the analytical evaluation (alcohol, sugar, SO_2 content, etc.) account for a maximum of 76 points.

483. Bespaloff, Alexis. **Guide to Inexpensive Wines**. New York, Simon and Schuster, 1973. 160p. $5.95. LC 72-93505.

Discussions and personal observations on retail wine prices, buying wine, and understanding the labels. Tasting is explained, along with storage and serving of wine. General redundant information includes characteristics of red, white, and rosé wines. Data about 2,000 specific wines that cost less than $3.50 per bottle (both imported and domestic) is still current, if one increases the ceiling of most of the imports by about a dollar. Fifty jug wines, available by the gallon, are also rated.

484. Blumberg, Robert S., and Hurst Hannum. **The Fine Wines of California**. Garden City, N.Y., Doubleday, 1971. 311p. illus. index. $6.95; $2.95pa. LC 72-131068.

The authors' purpose in this book is to provide the consumer with a way of differentiating among the many similarly priced wines available today. Unlike Massee's *McCall's Guide to Wines of America*, this book does not emphasize the vineyards and the backgrounds of California winemaking. It concentrates on "fairly detailed descriptions of well over 250 wines produced by the major California premium wineries . . ." (p.81). While only premium table wines from larger wineries are emphasized, Part Three describes some wines from smaller wineries and Part Four is a short description of bulk (jug) wines. Part Five describes wine labels, suggests books for further reading, and provides a list of wineries (addresses and phone numbers). The novice wine taster may enjoy comparing his evaluations with those of the authors, who are amateurs themselves writing a straightforward account for other amateurs. The authors visited 24 major wineries and 20 small ones.

485. Broadbent, J. M. **Wine Tasting: A Practical Handbook on Tasting and Tastings**. 2nd ed. London, Wine and Spirits Publ. Co.; distr. International Publications Service, 1971. 106p. $3.50.

Broadbent is a professional taster who now works for Christies. He gives the basics, the approach, the tasting expertise, and how to record one's impressions. Glossaries are appended. This book is almost as important as Marcus (see entry 488) and Max Lake's *The Flavour of Wine: A Quantitative Approach for the Serious Wine Taster* (Australia, Jacaranda, 1969).

486. Consumer Reports. **The Consumers Union Report on Wines and Spirits: Ratings, Recommendations and Buying Guidance on Table Wines, Sherry, Port, Champagne, Vermouth, Whiskies, Gins, Vodkas, Rums, Brandies, Cordials, Premixed Cocktails, and Cocktail Mixes**. Rev. ed. Mount Vernon, N.Y., Consumers Union, 1972. 216p. $2.00.

Changes since the 1963 edition were reflected in a CR series that ran from July 1969 through January 1972. New tastes meant new drinks, and thus changes of the past decade include the addition of sherry, port, champagne, vermouth, premixed cocktails, and cocktail mixes. The only beverage missing from this book is beer, adequately covered in the August 1969 issue (pp. 474-77). Apparently, cider, perry, and liqueurs are non-existent. Each section describes the latest techniques, production methods, merchandising, and pricing. Information is given on what to look for in tasting. Leading brands are rated in a blindfold test. Tennessee whiskies are included, but German Sekt is not. All brands of table wines tested were inexpensive at the time, but price rises have made the inevitable impact on those bottles in demand. In all cases, domestic bottlings were compared with foreign originals (e.g., wines, vodka) with surprising results.

487. Consumers' Association (U.K.). **The Which? Good Drink Guide**. London, 1970. 96p. illus. charts. £1.00.

These are tests of wines and spirits by a British consumer group. It is similar to CU in the United States, but its comments are more acerbic. Missing here are evaluations of beers, cocktails and liqueurs. This was a large-scale tasting with professionals and a computer program. Section one was "Red Wines," which included blends, cheap reds, Bordeaux, Burgundy, Nuits St. Georges, Beaujolais, Rhone, Italian, and Egri Bikaver. British shippers and British prices are given; thus, the book is of limited value in North America. However, the miscellaneous areas contain general information, including description of wine lists and import charges that are applicable to North America. Other sections deal with white wines, rosés, champagne and sparkling wines (such as German Sekt), fortified wines, and spirits (brandy, whiskey, gin, vodka, rum). In Britain most wines arrive in bulk and are bottled on the island. This is not so in North America except for Quebec. Superb tasting notes.

488. Marcus, Irving H. **How to Test and Improve Your Wine Judging Ability**. Berkeley, Calif., Wine Publications, 1972. 88p. illus. $2.25.

Twenty-five short chapters describe the basics and give tips. Marcus outlines the professional's approach to wine judging and describes the tasting tools and patterns. He gives a detailed explanation of the 20-point score card, and suggests ways of testing an individual's tasting ability. This former editor of *Wines and Vines* has devised a series of six sensory tests to measure taste thresholds for total wine evaluation. After describing the major components of wine, he analyzes the physiological capabilities and limitations of sight, smell, taste, and touch in judging wines. This is a superb and vital book in every way, and should be in every wine lover's collection.

489. Monmouth Wine Society. **Annual**. Fairhaven, N.J., 1972– .

As a typical example of what a club can produce, the M.W.S. put out their first annual for 1972 with an in-depth review of 300 wines. Also included are articles on wine buying, wine making, and food recipes.

490. Waugh, Harry. **Bacchus on the Wing: A Wine Merchant's Travelogue**. London, Wine and Spirit Publications, 1966. 203p. illus. $5.00. LC 67-74849.

491. Waugh, Harry. **The Changing Face of Wine: An Assessment of Some Current Vintages**. London, Wine and Spirit Publications, 1968. 109p. illus. $5.00. LC 74-401286.

492. Waugh, Harry. **Diary of a Wine Taster: Recent Tastings of French and California Wines**. New York, Quadrangle, 1972. 228p. illus. $6.95. LC 71-190483.

These three books are all very similar, having been written for the current times with up-to-date tastings in mind. They are systematic assessments of French wines (usually), district by district and year by year, with comparative data. Current vintages are stressed, and most of the material is derived from his writings for the British *Wine Magazine*. All are well illustrated, with photographs, maps, and charts. Waugh is a wine merchant, a consultant to the industry and a director of the Chateau Latour Vineyard. The *Diary* is written in that form, and covers his 1970 and 1971 trips to America. In this most recent book, he concentrates on Bordeaux, Burgundy, California reds, and a very few California whites. Comparisons are made between vintages, and between California and France. This is Waugh's first American appearance with this type of book; *Diary* was published in England as *Pick of the Bunch* in a slightly longer version.

493. **Wine Cellar Album: A Personal Record of Purchases and Usage of Wine for Your Greater Enjoyment of Nature's Unique Beverage**. San Francisco, Fortune House, 1970. unpaged. illus. $7.50. LC 72-138954.

Arranged in a looseleaf format (with room for extra pages at a later date), four sections cover appetizer wines; white, rosé and red wines; dessert wines; and sparkling wines. For each section there are purchase records on one side of the page (reference number, quantity, wine types, brand, purchase source, date, price, comments) and use records on the other side (reference number, wine type, number used, stock balance, rating, occasion, guests, food, comments). A miscellaneous section has guidelines for choosing glasses, suggestions for rating, starting a cellar, building a rack storage, glossary, bibliography, and space reserved for photographs and memorable labels.

494. **Wine Record Book**. 3rd ed. London, Wines and Spirits Publications, 1972. 144p. £4.40. ($12.00 U.S.).

In addition to a layout similar to that of *Wine Cellar Album* (above), this book contains hints on recording and rating wines, a glossary of tasting terms, a stock analysis, vintage and vinification charts, pages of descriptive matter for bottles, glasses, and openers, and an index page. Initials in gold can be added at no extra charge. Differences in publishers reflect differences in miscellaneous coverage. Actually, neither of these books is really needed. The wine lover need only keep a looseleaf notebook with the wine label information written down, plus whatever else he wishes to retain.

Chapter 8

PERIODICALS

GENERAL

There are only two magazines that will appeal to the novice or uninitiated layman—*Vintage* and *Wine Magazine*. However, the specialist would like to delve much deeper, and thus we have been generous in selecting periodicals for this section. Most general food magazines may include special sections on alcohol, and men's magazines (such as *Esquire*, *Penthouse*, or *Playboy*) certainly do. This section also contains a small section on newsletters.

As alcoholic beverages are consumed by a wide variety of people, written material on this form of enjoyment is not restricted to food and beverage magazines. Indeed, occasionally a very informative article appears in a general, regional, or special interest journal. For instance, by mid-1973 there had been very little written about the new "light" whiskies that had been on the market for a while, but *Time*, despite the attempts of distillers to disguise the lack of sales, came up with a negative story in the March 5, 1973, issue (p. 63). The following three items, taken from special interest magazines, are especially worth seeking out.

495. Burck, Gilbert. "Good Living Begins in the Wine Cellar." **Fortune**, June 1, 1967, pp. 122-29.

A story of the great private wine cellars in America today, illustrated with photographs of the contents and comments by their owners: Jules Hern, Stanley Marcus, and Ted Bensinger, among others.

496. Cady, Lew. "You Only Go Around Once in Life; Throw Away the Beer and Keep the Can." **Oui**, November 1972, pp. 75-77.

A report on Beer Can Collectors of America, illustrated with color photographs of cans. Apparently the only publication thus far where such photos can be seen.

497. **Consumer's Reports**, July 1969–January 1972.

Over this two and a half year period, CR reported on "Bourbon" (July 1969), "Beer" (August 1969), "Champagne" (November 1969), "Premixed Cocktails" (February 1970), "Cocktail Mixes" (September 1970), "Red Wines" (October 1971), "White Wines" (November 1971), and "Rosé Wines" (January 1972)–all from the point of view of consumer education, emphasizing method of production, merchandising, pricing, tasting, and "Best Buys."

REGIONAL PUBLICATIONS

Regional publications vary widely, but regional editions of nationally distributed magazines do produce a number of inserts that are pamphlet size or larger. These are extremely difficult to get hold of if you live outside the region, and are indexed only if the periodical received for indexing also contains the appropriate insert (which it will not in many cases, since indexing is often done in New York City instead of on a regional basis). Here are some examples:

498. **Financial Post Magazine** (Canada), December 1971, pp. 65-80 (16-page tear-out).

"Guide to Gourmet Living," which includes recipes and comments on wines available in Canada.

499. **New York Magazine**, February 2, 1972, pp. 43-78 (36-page tear-out).

"The Whole Grape Catalogue," which detailed the incestuous relationships among wine importers, wine merchants, and wine writers, commented on New York vineyards, described Alex Bespaloff's $1,000 wine cellar, discussed the best all-round buys in New York City (along with wine tastings), and concluded with short material on California wines. In what we hoped would be a series, Alexis Bespaloff reported on "The Great Beaujolais Tasting" for *New York Magazine* (January 29, 1973, pp. 62-67). Here was a general description of Beaujolais, a tasting of 149 different bottles available in New York City (from ordinary Beaujolais to a commune name), rated 1–20, and information on where to buy them at lowest prices. Because New York is America's largest city, it has the largest selection of Beaujolais in the country, hence this article would be ideal for the consumer who lives outside New York and has noticed, say, about two dozen different Beaujolais in his home town and wonders which one to try. Other magazines may have compared these wines on a national basis, but not all 149 types are distributed nationally, so the selection would be poorer. As it is, the reader can check to see which brands are available regionally and can compare with the *New York* article.

GENERAL PERIODICALS

500. **American Journal of Enology and Viticulture**. 1954– . American Society of Enologists, P.O. Box 411, Davis, Calif. 95616. Quarterly. $15.00 per year.

A professional journal which contains reports and reviews of original research on the grape vine and its products, with special emphasis on wine and brandy. Added features are bibliographic notes, abstracts, and charts. Indexed in: *Biological Abstracts* and *Chemical Abstracts*.

501. **Australian Wine, Brewing and Spirit Review**. 1882– . 13-31 Barrett Street, Kensington, Victoria, Australia. Monthly. $15.00U.S. per year.

Gives superb coverage of the "Down Under" wine industry, with charts and market patterns, plus profiles of the wine merchants and wine proprietors. Articles are on wineries, consumer consumption, equipment, markets, microbiology, and grapes. In content it is halfway between *Vintage* and *Wines and Vines*.

502. **Bordeaux Drinker's Companion**. 1972– . Box 2827, Los Angeles, Calif. 90028. Monthly. $7.50 per year.

This is a tasting guide for 10 to 14 Bordeaux wines rated monthly (great chateaux to cru bourgeois). Included also are brief technical descriptions and historical notes on a different estate in each issue. Posters of labels are available.

503. **Brewer's Digest**. 1926– . Siebel Publ. Company, 4049 West Peterson Avenue, Chicago, Ill. 60646. Monthly. $6.00 per year.

This is a technical publication, dealing with brewing, fermentation, malt, and so forth, with statistics and trade news. Very useful for announcements of new products, trends, history, book reviews, and so forth. *The Annual Buyer's Guide and Directory* is published as part of the January issue and is available separately at $2.00.

504. **Brewers Guardian**. 1871– . Northwood Publications, Ltd., Northwood House, 93-99 Goswell Road, London EC1V 7QA, England. Monthly. £5.00 per year.

This British equivalent to *Brewers Digest* is a trade publication, covering much the same as the latter, but with additional information on the European scene and some material on wines and spirits.

505. **The Compleat Imbiber**. 1956–1971. Published and edited by Cyril Ray. out of print.

These interesting and chatty newsletters were, for 15 years, personal statements about the wine and spirits trade in general. Beautifully designed, with colored illustrations much like those in children's books. Histories, recipes, and the lighter side of wine tasting and drinking were covered. In January 1974 this became a monthly column in *Wine Magazine.*

506. **Cuisine et Vins de France**. 1947– . 94, rue du Faubourg Saint-Honoré, Paris, 8e, France. Monthly. 40 Fr.Fr.(foreign rate) per year.

This is the French equivalent to the American *Gourmet*, although there is more emphasis on wines. As a publication in French for gourmets, amateurs, and wine connoisseurs, it presents menus and recipes complemented with the relevant wines. Regional items are displayed, along with articles on foreign cuisines. Studies are reported on for wines and the vine, and restaurants are rated.

507. **Gourmet: The Magazine of Good Living**. 1957– . 777 Third Avenue, New York, N.Y. 10017. Monthly. $6.00 per year; $18.00 for five years.

Contains many articles on foods that involve the use of wine or spirits. There is often an occasional article on wine, but little on spirits. The restaurants of London, New York, and Paris are reviewed. Travel articles give prominence to wine (e.g., a recent article dealt with the wine bars of London).

508. **Harpers Export Wine and Spirit Gazette**. 1952– . Harpers Trade Journals, Ltd., Southbank House, Black Prince Road, London SE 1, England. 3 issues a year (Feb./June/Oct.). $4.00 per year (foreign rate).

While the weekly (see entry 509) presents trade information, this valuable supplement, with texts variously in English, French, German, Portuguese, and Spanish, is crammed with articles and is geared to its foreign subscribers (hence the term "export" in its title). Feature stories cover one aspect of two or three countries (e.g., the February 1973 issue dealt in depth with Cognac).

509. **Harpers Wine and Spirit Gazette**. 1880– . Harpers Trade Journals, Ltd., Southbank House, Black Prince Road, London SE 1, England. Weekly. £8.00 per year (including Directory and Manual).

Trade news, market information and values, short concise statements of facts, statistics. For articles, see the Harpers Export issue (entry 508).

510. **O.I.V. Bulletin**. 1928– . Office International de la Vigne et du Vin, 11 rue Roquépine, Paris 8e, France. Monthly. 60 Fr.Fr. per year (foreign rate).

This is the leading wine organization of the world. This French language scholarly publication gives worldwide coverage to viticulture, winemaking, and wine economy. It details technological developments and also carries some general articles (such as "tourist influence on wine consumption"). Sections include notes from the world press, new legislation (worldwide), statistics, and abstracts of the major wine journals.

511. **Revue Belge des Vins et Spiritueux.** 1945– . 45 rue Defacque, 1050 Bruxelles, Belgium. Monthly. 440 Belge. Fr. per year (foreign rate).

This is one of the leading non-French wine periodicals (the Belgians consume an enormous quantity of wine, but do not grow any of their own). Language is French, with Flemish summaries. Foreign wine production is summarized; there are extensive sections dealing with French wines, aperitifs, auction notes and legislation; and there are cocktail and food recipes. The January 1973 issue contained an important editorial on rising wine prices.

512. **Rue du Vin de France**. 1927– . 94, rue du Faubourg Saint-Honoré, Paris 8e, France. 5 issues per year. 45 Fr.Fr. per year (foreign English edition).

This sister publication of the *O.I.V. Bulletin* is available in English translation, but it devotes all of its space to France. Each issue has quality documentation and 60 to 80 pages of text; it is illustrated. Type of material covered includes histories of wines and glasses, labels, the value of harvests, tasting notes, wine lists in restaurants, legislation, food, amateur wine makers, and book reviews.

513. **Revue Vinicole Internationale / International Wine Review: Revue des Vins, Vins de Liqueur et Apéritifs, Eaux-de-Vie de Liqueurs, Jus de Fruits, Cidres, et Boissons de Qualité**. 1880– . Compagnie Française d'Editions, 40 rue du Colisée, Paris, 8e, France. 8 issues per year. 70 Fr.Fr. per year (foreign rate).

With text in both French and English (plus appropriate translated summaries), this periodical covers technical, trade, and regional studies of French alcoholic beverages. There is a food section, news and statistics, bibliographies and book reviews.

514. **Vintage Magazine: A Way of Life**. 1971– . P.O. Box 866, New York, N.Y. 10010. Monthly. $12.00 per year.

This lavish publication grew out of the American Wine Society's newsletter, *Vintage* (edited by Phillip Seldon). Beginning with wines, it has now expanded to include spirits, liqueurs and food. In each issue there are *Gourmet*-type recipes and menus, plus recommended pre-packaged gourmet food products. Baron Roy de Groot has a food column. Unfortunately, all this leaves fewer pages for wine, and no mention of beer has yet appeared in

Vintage. Contents also include: wine news; advisory service; "Vintage Recommends" (tasting notes for wines priced under $5.00, with distributors and labels shown); vineyards, wineries, types of wines, geographical districts, countries (with maps and wine labels); "Workshop at the Waldorf Astoria," a panel discussion on wines and food; "The Enthusiastic Amateur," for novices; "Tasting Notes," from people who write in; "The Home Vintner," by authority Stanley Anderson; "Vintage Glossary," a pronunciation guide and definitions of terms used in that particular issue; and classified ads for swaps and sales. There are regional editions which vary only as to a four-page inset with wine ads, prices, and restaurants from local sources. The magazine has toned down since its beginning, for it rarely reviews books; it used to have commentaries on wine stores in New York City and Chicago (this was to be projected to include all major urban centers in the United States); commentaries on notable wine cellars in private hands; and individual sections for news and articles from Bordeaux, Burgundy, Germany, and California. These are all gone, having been replaced by food and gracious living articles. *Vintage*, instead of concentrating on publishing about wine, has diversified into a *Culinary Arts Society* (mostly a book club), another new promotional magazine called *Wine and Spirits* (available exclusively at selected retail wine outlets), an annual buying guide (originally scheduled for 1972, now postponed to 1974 because of rising wine costs), and a wine appreciation course at $65.00 for *Vintage* subscribers (Vintage Wine and Dine Society).

515. **Wine and Food: A Gastronomical Quarterly**. 1934– . (No. 149, June 1970). International Wine and Food Society, London, England. out of print.

This was similar to *Gourmet*, with more emphasis on wines. It was offered by the Society to its membership, which now receives *Wine Magazine*. *Wine and Food* was absorbed by the British *House and Garden*; it is now a monthly bound-in supplement of 20 pages, with news, tips, articles, etc. The annual quantity of text approximates the cumulative number of pages formerly published on a quarterly schedule.

516. **Wine Magazine**. 1958– . Wine and Spirits Publications, Ltd., Southbank House, Black Prince Road, Lambeth, London, SE 1, England. Monthly. $10.00 U.S. per year.

Originally a quarterly, this periodical is now sent monthly to members of the International Wine and Food Society. Articles deal with wine tours (mostly Europe), tasting notes, news, food, and consumer education. "Will You Dine Out with Our Master of Wine?" is a very critical, highly skilled write-up of a group dining and wining in a finer restaurant to establish the complementary relationship between wine and food. Although less compartmentalized than *Vintage*, its special section "News from Bordeaux" reflects the British interest in claret. Harry Waugh has a regular column, and there is an exceptional and controversial letters column. Not quite so lavish as *Vintage* (illustrations are mostly black and white), but it does have a regular section, "Bookshelf," in which printed materials are mentioned in passing, while *Vintage* rarely mentions new books.

517. **Wine Review**. 1964– . Southern Cross Books, 5 Pencarrow Avenue, Auckland 3, New Zealand. Quarterly. $2.00 NZ per year (foreign rate).

Very similar to the *Australian Wine, Brewing and Spirit Review*, with industrial news, foreign coverage, charts and market patterns, profiles of wine proprietors, wineries, consumer consumption, articles dealing with microbiology, grapes, and equipment. Book reviews are included.

518. **Wine Scene**. 1973– . Box 49338, Los Angeles, Calif. 90049. 10 issues per year. $10.00 per year.

Wine consultant John Morin edits this enchanting series of California wine tasting notes. Information usually consists of opinions, prices, sources, and new releases. Scope is limited to California wines.

519. **Wine Spirit Trade International**. Haymarket Publishing, Ltd., Gillow House, 5 Winsley Street, London, W1A 2HG. Monthly. £23.50 per year.

Trade news and markets, statistics, etc., similar to the information in *Wines and Vines* or in the two Harpers publications.

520. **Wines and Vines**. 1919– . 703 Market Street, San Francsico, Calif. 94103. Monthly (semi-monthly in December). $8.00 per year. Title varies: **California Grape Grower** (1919-1929); **California Grower** (1929-1933).

This is the American wine trade industry magazine, with comprehensive coverage of the U.S. production of wine through marketing statistics, charts, and news. Articles deal with proprietors and grapes. Foreign coverage is excellent, but the foreign wine trade is dealt with as though it were competition. Book reviews are included.

HOME WINEMAKING

Interest in home winemaking has exploded in the popular press. Articles have appeared regularly in Sunday supplements, although they are mostly of the "how I started" variety, without carrying through to a successful conclusion. Some articles of note over the past few years include "The Potable Hobby" (*Newsweek*, Sept. 14, 1970, p. 100); "Vin du

Pays–American Style" (*Holiday*, May 1970, pp. 60-61); and "I Made My Own Wine and Lived to Write About It" (*Esquire*, January 1974, pp. 122-23, 166-68). Interestingly enough, no equipment evaluations have yet appeared in *Consumers Report*. Special sections are reserved in *Vintage* and *Wines and Vines* for the amateur or home vintner, although there is no such section in *Wine Magazine*.

521. **Amateur Enologist**. 1970– . P.O. Box 2701, Vancouver 3, B.C. Quarterly. $2.00 per year.

Edited by Stanley F. Anderson, founder of Wine-Art, this quarterly publication contains a dozen excellent articles per issue on the subject of home wine- and beer-making. For example, the Spring 1973 issue contained an outstanding article on hybrid grapes for northern areas, another on coping with corks, and a third on barleys and malts for beermakers. This really is a Canadian oenologist's magazine since most, if not all, of the advertising is from Canadian suppliers. An occasional book review adds to its value.

522. **Amateur Winemaker**. 1957– . South Street, Andover (Hants.), England. Monthly. $5.00 per year.

This is the leading British publication; it includes recipes, articles, a queries column, and occasional book reviews. The publisher also issues many slim paperbacks of items culled from previous issues, such as *First Steps in Winemaking*, *130 New Winemaking Recipes*, and *Making Wines Like Those You Buy*. As with all British publications, sugar and Imperial measurements must be watched carefully. Yet 35,000 subscribers around the world cannot be wrong!

523. **Purple Thumb: Magazine for Home Beverage Makers**. 1966– . Wine Press Publ. Co., Box 2008, Van Nuys, Calif. 91405. Irreg. $1.00 per issue.

This periodical is unusual in that it was founded by vintners from Oregon, where it was initially issued. Since then, it has packed off for sunny California. It is very well laid out–perhaps the best designed of all such periodicals–with heavy emphasis on pictures and how-to articles. Some previous material has been gathered into its monograph, *Wine, Wisdom and Whimsey* ($5.95).

NEWSLETTERS

All of the following newsletters are available free upon request, provided that the publishers are not swamped with orders. Each, of course, emphasizes its own wine product, but such emphasis runs from "slight" to "heavy." Be aware that this is promotional literature and that any names on the mailing lists may be used for other commercial ventures. Explanatory notes are given where the newsletter diverges from strict promotionalism.

524. **Bottles and Bins**. Charles Krug, c/o C. Mondavi Sons, Box 191, St. Helena, Calif. 94574.

Published since 1949, the articles (including recipes) are written by Francis L. Gould.

525. **Brookside Wine Press**. Brookside Vineyard Company, Guasti Plaza, 9900 A Street, Guasti, Calif. 92743.

526. **From the Vineyards**. Almaden Vineyards, One Maritime Plaza, San Francisco, Calif. 94111.

Irregularly produced, this publication gives news of Almaden's products (in a non-commercial style), information on the grape types used, plus selected recipes from Bay Area restaurants.

527. **Heitz Cellar Newsletter**. (Heitz Wine Cellar, 500 Taplin Road, St. Helena, Calif. 94574.

528. **Inglenook News**. The Cellarmaster, Inglenook Vineyards, Rutherford, Calif. 94573.

Items include informative advertisement reproductions, suitable for framing.

529. **Mirassou Latest Press**. Mirassou, Route 3, Box 344, San Jose, Calif. 95121.

530. **Sebastiani Newsletter**. Sebastiani Vineyards, P.O. Box AA, Sonoma, Calif. 95476.

This monthly newsletter gives up-to-date information on the climatic conditions in the Sonoma Valley. In addition, progress reports are made on Sebastiani wines (both in the cask and in sales), and articles are written on technical processes (e.g., pruning techniques) as used by the vineyard owners.

531. **Tiburon News**. Tiburon Vineyards, Route 3, Box 344, San Jose, Calif. 95121.

532. **Vignettes de Vignobles**. Frederick Wildman and Sons, 21 East 69th Street, New York, N.Y.

A quarterly (seasonal) comment on how the European wines are maturing; the fall and winter issues contain information on current crops and harvests. One of the few newsletters put out by a wine importer, and the least commercial.

CATALOGS

Every store or importer may have a price list for his offerings, but a few have catalogs that a consumer may wish to refer to from time to time. As a judicious selection, the following are presented, two from auction firms and two from wine merchants. Sotheby's of England (34-35 New Bond Street, London W1A 2AA) also offers catalogs, but these must be ordered individually at $0.75 each *three* weeks before the sales. As these are not kept by Sotheby's, the hobbyist must know when the sales are to occur. Only two merchants are included here because they are the only ones that responded to letters; by extension, they are obviously the only ones that would respond to your letters. A drop-in visit to the stores of other wine merchants may produce more favorable results.

533. Berenson, Ltd. (Boston). **Collection of Fine Wines.** Annual.

This free catalog is one of the better ones put out by a retailer. It is well put together, with geographic descriptions, vintage charts, wine futures, histories, and maps. The wines come from all over the globe, but nevertheless there is stress on French and German wines. Some important spirits are also listed; there is guidance for food and wine partnerships. Illustrations include labels and photographs of towns and vineyards.

534. **Christie's Wine Review.** 1972– . Christie, Manson and Woods, 867 Madison Avenue, New York, N.Y. 10021. Annual. $4.50.

The 1972 issue had about 2,000 price entries for the wines auctioned off in 1971. Articles included Edmund Penning-Rowsell on the history of Christie's from 1766 to 1800 and on the wine market of today; Harry Waugh's personal assessment of the 1960s Bordeaux; and Michael Broadbent on the original rating and present drinking conditions of older vintages. The 1973 issue of 72 pages covered 2,330 price entries (e.g., Bordeaux: 200 chateaux, 1846 to 1969; 57 shippers, 1834 to 1967; brandy: 100 types produced since 1789), and there were articles by Edmund Penning-Rowsell on current wine markets and the surging prices; his continuing history of Christie's (1801 to 1850); Cyril Ray on Chateau Lafite; Archie Ling on the passion for wine trade relics and collector's items; and Michael Broadbent on port and 150 years of port vintages.

535. Heublein, Inc. **National Auction of Rare Wines.** Box 956, Hartford, Conn. 06101. Annual $2.50.

Heublein, in addition to being a producer of wines and spirits, is one of the two great wine auction firms. Although a distiller, it has an annual sale every May 31, dealing mainly with rare or unusual wines, and many from California's historical days. This catalog has all the listings with some illustrations and notes, plus a comparative review of wine prices in both Europe and the United States.

536. House of Hallgarten. **List of Fine Wines, Spirits and Liqueurs.** Carker's Lane, Highgate Road, London NW5 1RR. Annual. $1.00.

This catalog is a guide to several hundred wines (mostly German), with maps, tasting notes, vintage reports and vintage charts, and, of course, prices.

Chapter 9

ASSOCIATIONS AND CLUBS

ASSOCIATIONS AND TRADE GROUPS

These associations and trade groups are mainly promotional. There are several closed societies that serve members only, but as they deal with a service, some are happy to give out data and information on request. Some others have requested that we not mention their names, and these requests have been honored. For information about other countries (such as Argentina, Yugoslavia, or Japan), the best recourse is to their Trade Consulate or Ambassador's Office.

AMERICA

Over the 1973/74 period, American liquor associations are being merged into the Distilled Spirits Council of the United States, Inc. As announced on March 23, 1973, DISCUS will be "the national trade association of the domestic distilling industry," and the Bourbon Institute, Distilled Spirits Institute, and the Licensed Beverage Industries will all be consolidated into operating divisions. Information can still be obtained from these three associations as DISCUS will be working on a national level primarily as a consolidated lobby group; however, matters regarding co-ordination should be addressed to DISCUS.

American Society of Barmasters. P.O. Box 1080, Louisville, Kentucky 40201.

"To stimulate increased public acceptance of our nation's fine bars and taverns." It encourages the creation of original mixed drink recipes through its biennial Early Times National Mixed Drink Competition. It is a source for cocktail recipes.

American Society of Enologists. P.O. Box 411, Davis, Calif. 95616.

"To promote technical advancement of enology and viticulture through integrated research by science and industry; to provide a medium for the exchange of technical information; to improve wine and grape quality." Publishes the monthly *American Journal of Enology and Viticulture*.

Bordeaux/Alsace Information Bureau. c/o Bell and Stanton, 909 Third Avenue, New York, N.Y. 10022.

As an office for potential tourists to France, this center provides a wealth of travel information from a variety of sources. In addition, it can also provide wine posters, charts, vineyard maps and labels (either free or for a small sum).

Bourbon Institute. 277 Park Avenue, New York, N.Y. 10017.

Now an operating division of DISCUS. Promotes the sale of bourbon through public relations activities and government liaison. Publishes *Facts About Bourbon* (free. 24p.) plus various cookery pamphelts, *Book of Bourbon*, and so forth.

California Brandy Advisory Board. 414 Jackson Square, San Francisco, Calif. 94111.

A trade group concerning the promotion of California-produced brandy. Publishes a free, 16-page collection of recipes.

Distilled Spirits Council of the United States, Inc. (DISCUS). Pennsylvania Building, Washington, D.C. 20004.

This is now the national trade association of the domestic distilling industry; it works with the public through its three operating divisions: Bourbon Institute, Distilled Spirits Institute, and Licensed Beverage Industries.

Distilled Spirits Institute. 1132 Pennsylvania Building, Washington, D.C. 20004.

Now an operating division of DISCUS. It provides statistical and legal information for the industry and the public. Maintains a library, publishes a biennial *Summary of State Laws and Regulations Relating to Distilled Spirits*, the *Annual Statistical Review of the Distilling Industry*, and various pamphlets dealing with legislation and local options, statements on positions adopted regarding pollution, waste, prices, and bottle stamps.

Food from France, Inc. Food and Wine Information Center, 1350 Avenue of the Americas, New York, N.Y. 10019.

Represents various trade and promotional groups dealing in French wines, food, and tours. A one-stop place of inquiry for the novice, although he will not get the addresses supplied below in the section on France.

German Wine Information Bureau. 666 Fifth Avenue, New York, N.Y. 10019.

Representing the leading wine promotion groups in Germany, it publishes the newsletter "Wineland Grapevine," prepared by the German Wine Exporters Association in Germany. This deals with vintage reports and correct growing conditions.

Greek Trade Center. 150 East 58th Street, New York, N.Y.

As an office for potential tourists to Greece, this center provides a wealth of travel information from a variety of sources. In addition, either free of cost or for a small sum, it can provide wine posters, charts, vineyard maps, and labels.

Joint Committee of the States to Study Alcoholic Beverage Laws. 5454 Wisconsin Avenue, N.W., Suite 1610, Washington, D.C. 20015.

Undertakes studies for state liquor control agencies. It attempts to bring about a uniform approach to solutions of common problems. Studies have included: trade barriers affecting interstate commerce; sales to minors; mark-up policies and subsequent sales impact; uniform standards for advertising.

Licensed Beverage Industries. 485 Lexington Avenue, New York, N.Y. 10017.

Now an operating division of DISCUS, its purpose is "to deal with all matters and problems affecting public attitudes toward the industry and its products." Areas of interest include economic research, women, statistics. Maintains a library. Free pamphlets include: *L.B.I.: The Industry Forum and the Voice* (20p.); *An Old American Custom: Some Facts about Beverage Control in America* (16p.); *The A.B.C.'s of Beverage Control* (16p.); and *Moonshine: Misery for Sale* (24p.). The Women's Division puts out home entertainment materials, including the package kit for *Celebrate Together*, an audiovisual program on food and drink and fashion, plus recipes. Other working units, now also part of DISCUS, include: Tax Council of the Alcoholic Beverage Industry; National Council Against Illegal Liquor; and Women's Association of Allied Beverage Industries.

Master Brewers Association of America. 4513 Vernon Blvd., Madison, Wisc. 53705.

A professional association, open to brewers only; publishes the informative *MBAA Technical Quarterly*.

National Association of Alcoholic Beverage Importers. 1025 Vermont Avenue, N.W., Suite 1205, Washington, D.C. 20005.

Compiles and reports statistics (annual *Statistical Review*), and sponsors lectures at hotel and restaurant schools. A good source for statistics.

Portuguese Information Bureau. 727 Park Avenue, New York, N.Y. 10021.

As an office for potential tourists to Portugal, this center provides a wealth of travel information from a variety of sources. In addition, it can provide wine posters, charts, vineyard maps, and labels (either free or for a small sum).

Sommelier Society of America. 121 West 45th Street, New York, N.Y. 10036.

Its aim is "to impart greater knowledge of wines and spirits among those who serve and sell these products." Holds monthly tastings for members, plus correspondence courses. Maintains a library.

United States Brewers Association. 1750 K Street, N.W., Washington, D.C. 20006.

Represents manufacturers responsible for 86 percent of industry sales. Maintains a library of over 5,000 volumes on beer and the brewing industry, plus works providing information on products, marketing, employee relations, history, and Prohibition. Publishes *The Brewers Almanac* (annual) and various pamphlets such as *The Story of Beer: The Beverage of Moderation* (12p.), *Barley, Hops and History* (28p.), and materials for parties and cooking with beer.

Wine Advisory Board. 717 Market Street, San Francisco, Calif. 94103.

This is a promotional agency responsible to the California government for such promotion; it acts on behalf of the California wine producers. It offers a free correspondence course called "Wine Study," which offers a diploma upon completion. Among its many pamphlets are *An Introduction to Wine* (24p.), *Wine Tasting Party* (24p.), and *How to Cook with California Wines* (10p., 81 recipes). Its cookbooks are listed in the cookery section of this book. The Board also has a catalog of over 60 wine-related items offered for sale (such as approved all-purpose wine glasses, corkscrews, placemats to contain wine glasses with illustrations for tasting, etc.), and it promotes the *Wine Cellar Album* (see under Wines, above).

Wine Institute. 717 Market Street, San Francisco, Calif. 94103.

Founded in 1934 to upgrade the industry after Repeal, this association represents California wine growers; it conducts research and market surveys, directs public relations, sponsors wine technology studies, compiles statistics, and engages in wine history. It maintains a library. Its *Bulletin* contains annual reports in each June 15 issue, plus the appropriate statistics. It does promotion under contract to the Wine Advisory Board. Its many activities are performed with Board funds. One unusual feature of the Institute is that it provides *free* technical review service for checking the accuracy of any wine writer's manuscript.

CANADA

Brewers Association of Canada. Suite 805, 151 Sparks Street, Ottawa, K1P 5E3.

This is the national trade association of Canada's brewing industry. There are many pamphlets and much promotional material, most of which deals with cooking and beer. It collects and distributes industry statistics on sales and materials, performs research work, and maintains a library.

Canadian Committee of French Wines (SOPEXA). P.O. Box 177, Place Bonaventure, Montreal, P.Q.

Represents the French wine trade in Canada, but also handles American requests because it is well-equipped to do so.

Canadian Wine Institute. Room 800, 111 Richmond Street, Toronto.

Represents the interests of the wine industry of Canada, primarily the Niagara, Ontario, region. Maintains a library and publishes various pamphlets concerning recipes, tastings, and parties.

EUROPE

The wine trade associations in France and Italy have very heavy responsibilities. In addition to ensuring that minor regulations are carried out (and performing policing duties), they are also responsible for economic and statistical surveys. Their second large responsibility is for permanent contact between the growers and the merchants, much needed by the wine brokers. Their third responsibility (and the one of great interest to the reader) is the promotion of a demand for their products.

France

The French wine trade is very well organized in its approach to promotion—much more so than Italy, Germany, or Spain. Each wine-producing area has either a promotion board or a consortium that regulates production and engages in promotion. Various "comités de vin" are co-ordinated by the Comité National des Vins de France, and all have color maps of "Routes du Vin," usually 24" x 36", with French, English, and sometimes German texts. These texts concisely summarize the characteristics of the local wine. Highway maps are sometimes provided for tours, and these are detailed as to locations of vineyards. Cooperatives, especially in non-A.O.C. areas, carry out wide promotional and tourist activities, with statistics, brochures, more maps, histories, lists of restaurants and hotels, lists of exporters and importers, and so forth. No one printed book has all this information about the wine-producing areas of France, yet it is vital information for either the tourist or the armchair traveller. No matter what quality wines are produced in an area (and in some places no wines at all are produced), there is a local tourist office called the Syndicat d'Initiative. By writing to this office ahead of time, the traveller can receive much valuable information on wine (if applicable), restaurants, hotels, maps, and so forth. And all of it is free. To our knowledge, the following complete list appears in print for the first time. While it is alphabetical by association name, wherever applicable, the geographic region is in boldface type so that it stands out. Also, special publications beyond those listed above are noted.

Association de Propagande pour le Vin. 18 rue du 4 Septembre, 34 **Beziers**.

Bureau National Interprofessionnel de **l'Armagnac**, 32 Eauze.

Bureau National Interprofessionnel du **Cognac**. 3 rue Georges-Briand, 16 Cognac.

Comité Interprofessionnel de la Côte d'Or et de l'Yonne pour les Vins d'Appellation d'Origine Contrôllée de **Bourgogne**. Petite Place Carnot, 21 Beaune.

Also publishes a large book, *Vines and Wines of Burgundy*.

Comité Interprofessionnel de Saône-et-Loire pour les Vins d'Appellation d'Origine Contrôllée de Bourgogne et de **Macon**. 3 bis, rue Gambetta, 71 Macon.

Comité Interprofessionnel des Vins à Appellation Contrôllée de **Touranne**. Chambre de Commerce, 12 rue Berthelot, 37 Tours.

Represents the following wines: Vouvray, Chinon, Bourgueil, and Montlouis.

Comité Interprofessionnel des Vins **Côtes de Provence**. 3 avenue Jean-Jaurès, 83 Les Arcs-sur-Argens (Var).

Comité Interprofessionnel des Vins de **Gaillac**. 19 rue du Père-Gibrat, 81 Gaillac.

Comité Interprofessionnel des Vins des **Côtes du Rhone**. Maison du Tourisme et du Vin, 41 Cours Jean-Jaurès, 84 Vignon (Vaucluse).

Comité Interprofessionnel des Vins d'Origine du Pays **Nantais**. 17 rue des Etats, 44 Nantes.

Comité Interprofessionnel des Vins Doux Naturels et Vins de Liqueur à Appellation Contrôllée. 19 avenue de Grande-Bretagne, 66 Perpignan.

Comité Interprofessionnel du Vin **d'Alsace**. 8, Place de Lattre-de-Tassigny, 68 Colmar (Haut-Rhin).

Comité Interprofessionnel du Vin de **Champagne**. 5 rue Henri Martin, 51321 Epernay.

Publications include a wide variety of pamphlets, such as ones detailing what the Comité does, its setting of standards, declaration of vintage, etc.; another two on a basic description of champagne manufacture, storing and serving (*Well, There It Is* and *Choosing and Serving Champagne*); and the complete *Les Expeditions de Champagne, 1972*, which gives statistical breakdowns of the harvest. All of the above are mentioned because this is the most tightly controlled wine-producing area in the world.

Comité National des Vins de France. 43, rue de Naples, 75 Paris, 8e.

This umbrella organization forwards requests for information to the relevant body. It works with SOPEXA and Foods from France, Inc., and other trade bodies in other countries.

Comité National du **Pineau des Charentes**. 31, avenue Victor-Hugo, 16 Cognac.

Confédération Nationale des Industries et Commerces en Gros des Vins, Cidres, Sirops, Spiritueux et Liqueurs de France (C.N.V.S.). 103 blvd. Haussmann, 75 Paris, 8e.

Confédération Générale des Vignerons du **Midi**, 1 rue Marcelin-Coural, 11 Narbonne.

Conseil Interprofessionnel des Vins **d'Anjou** et de **Saumur**. 21 blvd. Foch, 49 Angers.

Conseil Interprofessionnel des Vins de **Fitou, Corbières, Minervois**. 55, avenue Georges-Clemenceau, 11 Lézignan-Corbières.

Conseil Interprofessionnel des Vins de la Région de **Bergerac**. Place du Docteur-Cayla, 24 Bergerac.

Conseil Interprofessionnel du Vin de **Bordeaux**. 1 Cours du XXX Juillet, 33 Bordeaux.

Fédération des Unions Viticoles du Centre le **Prieuré**. 18 Ménetou-Salon.

Fédération Régionale des V.D.Q.S. **Savoie-Bugey-Dauphine**. 2 Place du Château, 73 Chambéry.

Groupement de Développement Viticole. Maison de l'Agricluture, 46 avenue Jean-Jaurès, 18 **Bourges**.

Groupement d'Etudes Techniques Viticoles et Arboricoles de l'Yonne. Chambre d'Agriculture. 2 bis, blvd. Darout, 89 Auxerre.

Groupement Interprofessionnel des Vins de l'Isle de **Corse**. Centre Administratif, 20 Bastia, Corsica.

Maison du Vin. **Barsac** (Bordeaux).

Maison du Vin. **Blaye** (Bordeaux).

Maison du Vin. **Cadillac** (Bordeaux).

Maison du Vin. **St. Emilion** (Bordeaux).

Société de Viticulture du **Jura**. Maison de l'Agriculture, 39 Louis-le-Saunier.

Syndicat de Défense des Côteaux d'Aix-en-**Provence**. B.P. 38, 13 Aix-en-Provence.

Syndicat de Défense des Vins de **Madiran** et du **Pacherenc** du Vic Bihl. Château de Crouseilles, 64 Crouseilles.

Syndicat des Producteurs des Vins à Appellation Contrôllée **Jurançon**. 5 Place de la République, 64 Pau.

Syndicat du Commerce d'Exportation des Vins, Cidres, Spiritueux et Liqueurs de France. 103 blvd. Haussmann, 75 Paris.

Syndicat Régional des Vins de **Savoie**. 11, rue Métropole, 73 Chambéry.

Union Interprofessionnelle des Vins du **Beaujolais**. 210, blvd. Vermorel, 69 Villefranche-sur-Saône (Rhone).

Union Viticole Sancerroise, Comité de Propagande des Vins de **Sancerre**. Mairie, 18 Sancerre.

Vignerie Royale du **Jurançon**. 72, rue Castetnan, 64 Pau.

Great Britain

Only Scotch is very prominent here (England has little native wine). But the English, with their propensity for clubs, have created a number of groups for the pursuit of leisure activities (see under Clubs and Societies, below).

English Vineyards Association. York House, 199 Westminster Bridge Road, London SE 1.

Devoted to maintaining grape vineyards in England, it publishes a quarterly *Journal* for the dissemination of information. The climate in the United Kingdom is rarely suitable for grape wine; however, a large quantity of fruit wine is made.

Scotch Whisky Association. 17 Half Moon Street, London, W1Y FRB.

Designed "to promote and protect the interests of the Scotch Whisky trade at home and abroad." Produces movies and a Scotch Whisky Distillery map, statistical reports, and brochures such as *Scotch Whisky* (well illustrated with historical photographs), which describes the process of making Scotch plus questions and answers. Everything is available in six languages.

Wine Development Board. 6 Snowhill, London EC1A 2DH.

"To stimulate knowledge and appreciation of wine by increasing consumption of foreign wine in the United Kingdom." It publishes numerous booklets, numerous cookery recipes, tips on entertaining, etc. Its monthly newsletter is *Grape Vine*, and it is possible for a North American to get on the mailing list. The Board sponsors wine competitions and tastings, and it provides consumer education programs through lectures.

Italy

Tourist trade and information requests are usually handled by the cooperatives themselves rather than by a tourist office. Hence, material here tends to be more commercial–i.e., catalogs and labels rather than maps and general information. As the wines of Italy do not get the same veneration as the quality wines of France, the stress on finished product may be all that is needed here.

Comitato Nazionale per la Difesa delle Denominazioni di Origine dei Vini. Via Nizza 45, 00198 Roma.

A regulatory body–one of many–to protect the 1963 D.O.C. laws.

Consorzio dei Vini Tipici Piemontesi. Corso Alfieri 313, 14100 Asti.

Consorzio del Vino Asti ed **Asti Spumanté**. Piazza Roma 5, 14100 Asti.

Consorzio del Vino **Chianti** "Gallo." Via Valfonda 9, 50123 Firenze.

Consorzio del Vino **Chianti** "Putto." Piazza S. Firenze 3, 50122 Firenze.

Consorzio della **Grappia** Piemonte. Corso Alfieri 313, 14100 Asti.

Federazione Italiana Industriali Produttori Esportatori ed Importatori di Vini, Acquaviti, Liquori, Sciroppi, Aceti, ed Affini (Federvini).

An umbrella organization that will provide contacts. Publishes a weekly informative newspaper, *La Gazzetta Vinicola*.

Irvam. Via Castelfidardo 43, 00185 Roma.

Istituto Nazionale per il Commercio Estero. Via Liszt 21, 00100 Roma.

Istituto **Siciliano** della Vite e del Vino. Via Libertà 66, Palermo, Sicilia.

Regione Antonoma Trentino Alto Adige. Palazzo Regione, 30100 Trento.

Unione Italiana Vini. Via S. Vittore al Teatro 1, 20123 Milano.

Technical research, machines, and management affiliations. It publishes a weekly newspaper, *Il Corriere Vinicolo*, with a gorgeous quarterly color supplement (promotional pictures) called *Enotria*.

CLUBS AND SOCIETIES

Only national and international societies are listed here. Regional societies are too numerous, too varied, and of an uneven quality, and many do not accept members outside a definite geographic area. The best source for obtaining a list of these nearby clubs would be the largest retail wine and liquor merchant in town. Most publish a newsletter for membership information, and one of the best (Monmouth) is to be found in this listing as illustrative of what a regional club might do. Similarly, the International Wine and Food Society has a large listing of regional branches that are affiliates, and this information may be obtained by writing to the society. Some commercial (i.e., profit-making) clubs flourish, but they are best to be avoided because of the "Wines Internationale" fiasco. Most of the clubs deal with importing wine cases in bulk for resale to their "members" and are in effect simply retail wine importers doing business by mail. Of the more general societies dealing with wine, there are Les Amis du Vin (4701 Willard Avenue, Washington, D.C. 20015) and Vintage Wine and Dine Society (for subscribers to *Vintage* magazine).

The following clubs are more concerned with the "hobby" line than with the "appreciation" of wine, beers, or spirits. This is a logical extension to the routine consumption of alcoholic beverages, especially as these clubs utilize the service and containers of such beverages.

Antique Bottle Collectors Association. 1959– . P.O. Box 467, Sacramento, Calif. 95802.

Members (about 1,500, with 15 local clubs) include families and individuals who collect old or antique bottles, such as historical flasks, whiskey and liquor bottles, and so forth. Through a monthly publication, *The Pontil*, it promotes bottle collection and historical research into manufacturing methods. Annual meeting is in June in Sacramento.

Beer Can Collectors of America. 1970– . P.O. Box 9104, St. Louis, Missouri 63117.

The approximately 400 members collect beer cans (first produced in 1935) of rare types of beer that are no longer produced; these are usually regional varieties. Some members have over 5,000 cans. The prime purpose of the club is to provide members with an opportunity to trade beer cans, which is done at the club's travelling annual convention. Individual members can apply for a mailing list of collectors so that contacts can be made person-to-person rather than through the club. Annual *Can*vention is usually in the fall (held in Cincinnati in 1973).

British Beer Mat Collectors' Society. 1960– . 142 Leicester Street, Wolverhampton, WV6 0PS, England.

About 400 members participate in the hobby of drip-mat collecting throughout the world. Although British mats (introduced in 1920) predominate through breweries' monthly offerings, foreign mats are also collected. The 14 branches support a section: Beer Mat Museum, 6 Brackley Road, Beckenham, Kent, BR3 1RG, England. The fascinating *Beer Mat Magazine* is a monthly that lists items for trade and establishes the bid-ask prices for rare, mint-condition mats.

International Beer Tasting Society. 1956– . 801 Lido Sound, Newport Beach, Calif. 92660.

For "beer lovers who want to enjoy beer in the company of other beer lovers," this very loosely run club tries to foster appreciation of the art of brewing, and sometimes awards letters of commendation to breweries. Its members are to "grade and savor as many of the world's beers as possible, both as travelling individuals and as a formal beer tasting organization." No newsletter or other evaluative communication.

International Wine and Food Society, Ltd. 1933– . 44 Edgemore Road, London W.2, England.

Seven thousand individuals affiliated with 27 branches in Great Britain and 84 overseas promote the appreciation of wine and food, and seek improvements in standards of cookery. Their original publication, *Wine and Food*, has been subsumed, and now *Wine Magazine* is sent to members. Branches put out their own communications. The Society publishes many books, some of which are available in North America from commercial publishers. It maintains a library.

Labologists Society. 1958– . 335 Ditchling Road, Brighton, BN1 6JJ, England.

Two hundred members collect beer labels and study the history of breweries and the social history connected with the brewing trade. No formal communications with members.

Monmouth Wine Society. 1970– . 57 Heights Terrace, Fairhaven, N.J. 07701.

One of the better regional wine clubs with a first-class publication program. Family membership is $3.50 a year, which includes a monthly newsletter *plus* an annual. The 1972 annual reviewed 300 wines in depth, and also contained articles on wine buying and wine making, plus a collection of fine recipes. Annual is available separately at $3.50.

National Association Breweriana Advertising. 1972– . c/o *Brewers Digest*, 4049 West Peterson Avenue, Chicago, Ill. 60646.

Composed of people from throughout the United States who specialize in collecting breweriana, and for whom the travelling annual meeting is an occasion for trading or purchasing items as well as sharing information. No formal communications yet.

Society of Medical Friends of Wine. 1939– . Box 218, Sausalito, Calif. 94965.

Restricted to those physicians and surgeons interested in the nutritional and therapeutic values of wine. One of their purposes is to stimulate scientific research on wine; another is to develop an understanding of its beneficial effects and a third to encourage an "appreciation of the conviviality and good fellowship that are part of the relaxed and deliberate manner of living that follows its proper use." Quarterly dinners and annual vintage tour are highlighted for the 320 members who, not so surprisingly, live in the Bay Area. Their semi-annual *Bulletin* has articles, quotations, book news, and tasting notes.

Wine Label Circle. 1952– . Stadhampton Vicarage, Oxford OX9 7TU, England.

One hundred and fifty collectors of wine bottle labels meet in four branches to discuss the "stimulation and encouragement of research, the interchange of views between members, and the general study of decanter labels." A semi-annual *Journal* is somewhat like the *Beer Mats Magazine*, and this club resembles the Labologist Society, described above.

Chapter 10

MUSEUMS, LIBRARIES, AND CONTACTS

MUSEUMS

The traveller, either in Europe or at home, has ample opportunity to visit wineries (see Adams' *The Wines of America*, entry 213, and the list of foreign associations). Most wineries have a tasting room, which usually has a display of enchanting collections of materials from the past—perhaps an old bottle still containing the original wine, or tools or drinking vessels, even a cask or two. Most wine regions in Europe may put together a collection of equipment, art work, original tools, or mechanical devices, stemware and drinking vessels. Admission is often free.

AMERICA

San Francisco Wine Museum. Christian Brothers Rare Art Collection. Beach and Hyde Streets, San Francisco, Calif. Closed Mondays.

The only public museum in the United States exclusively devoted to presenting the story of wine and civilization through works of art, sculpture, artifacts, drinking vessels, and rare books. This permanent collection is based on four themes: 1) The Vine, the Grape and Harvest; 2) The Vintner and His Craft; 3) Wine Lore and Its Relation to Mythology; 4) The Celebration of Wine. There are more than 1,000 items here, dating from the Romans to the present time. Before the establishment of this new permanent "home," parts of the collection, generously donated by the Christian Brothers Winery, had been exhibited at Canadian and American art museums. A series of programs and special exhibits centering around this collection are also offered.

Greyton H. Taylor Wine Museum. Bully Hill Road, R.D.2, Hammondsport, N.Y. 14840. Open May 1 to October 31.

This museum housed the original Taylor Wine Company from 1883 to 1920. Included among its exhibits are rare items from Prohibition to Repeal, a large area devoted to the making of champagne in the nineteenth century, plus many displays centering around ancient presses and other old equipment and tools. In addition to a good wine and viticulture library, there is also a unique "Library of Grapes" from all over the world. These 200 grape varieties are kept flourishing in a temperature- and humidity-controlled glass enclosure.

EUROPE

Hospices de Beaune. Beaune, France. Open all year.

This group of buildings, which for 500 years functioned as a hospital, is now the site of an annual wine auction fund-raising drive to support current medical activities. Throughout the grounds are scattered many reminders of the past; a wine museum shows traditional equipment and methods of vinification.

Enoteca Italica Permanente. Sienna, Tuscany, Italy. Open all year.

This "wine library" displays the finest Italian wines available–both current and historical in production. All the wines may be tasted, either in the many tasting rooms or on the gazebo-like terraces where the vines are specially cultivated. Included in these many areas are interesting artifacts, equipment, and drinking vessels.

Palace of Wine. Jerez de la Frontera, Spain. Open all year.

This building was recently restored by the sherry producers who now use it as their headquarters. The traditional style of the building nicely complements the displays related to the making and enjoyment of sherry. This is the place to visit for information on the local *bodegas* and maps.

Villafranca Wine Museum. Villafranca, Panadés, Spain. Open all year.

Located in the ancient palace of the former kings of Aragon, this museum contains a collection devoted to the history of Spanish winemaking. The castle itself was built in 1285.

LIBRARIES

Generally, the academic and public libraries listed here will respond to requests from an applying local library to borrow books. In special circumstances, the other libraries (those attached to a business or research firm) will do likewise. In addition, many of the associations or clubs listed above also have working collections of books; because these are often not operating libraries, however, materials may not be borrowed. Nevertheless, personal correspondence or personal visits can sometimes provide access to the books; it does not hurt to try.

GENERAL

John Crerar Library. Chicago, Illinois.

Library of Congress, Washington, D.C.

National Agriculture Library. Washington, D.C.

New York Public Library.

U.S. Department of Commerce. Library. Washington, D.C.

U.S. Internal Revenue Service. Library. Washington, D.C.

University of California Library. Berkeley, California.

University of California Library. Los Angeles, California.

WINES

Canadian Wine Institute Library, Room 800, 111 Richmond Street, Toronto, Canada.

Brandy Wine Library, Fresno State College, Fresno, California.

Greyton H. Taylor Wine Museum. Bully Hill Road, RD 2, Hammondsport, New York 14840.

International Wine and Food Society, Ltd., Library. 44 Edgemore Road, London W2, England.

Napa Valley Wine Library. St. Helena, California.

New York Agricultural Experiment Station. Geneva, New York.

San Francisco Wine Museum. Library. Beach and Hyde Streets. San Francisco, California.

Sommelier Society of America. 121 West 45th Street, New York, New York 10036.

University of California Library. Davis Campus, Davis, California.
(In late fall, 1973, Maynard A. Amerine donated 2,000 English language books and 1,000 French and Italian books on the subject of wine to the Davis Library.)

Wine Institute Library. 717 Market Street, San Francisco, California 94103.

BEER

Anheuser-Busch, Inc. Library. St. Louis, Missouri.

Brewers' Association of Canada. Suite 805, 151 Sparks Street, Ottawa, Canada K1P 5E3.

Cincinnati Public Library.

Falstaff Brewing Corporation. Library. St. Louis, Missouri.

Fleischman Malting Co. Library. Chicago, Illinois.

Fleischman Research Laboratory. Library. Stamford, Connecticut.

Institute of Brewing Library. London, England.

Institute of Brewing Research Library. Nutfield, Surrey, England.

Schlitz (Jos. E.) Brewing Co. Library. Milwaukee, Wisconsin.

Seibel Institute Library. Chicago, Illinois.

U.S. Brewers' Association. Library. 1750 K Street, N.W., Washington, D.C. 20006.

SPIRITS

Bourbon Institute. Library. 277 Park Avenue, New York, New York 10017.

Distilled Spirits Institute. Library. 1132 Pennsylvania Building, Washington, D.C. 20004.

Filson Club Library. Louisville, Kentucky.

Licensed Beverage Bureau Library. 485 Lexington Avenue, New York, New York 10017.

Schenley Distillers Library. New York, New York.

CONTACT LISTS

Many of the associations, clubs and societies listed above (and particularly the latter two) can make suggestions or recommendations on almost every aspect of alcoholic beverages. Some clubs are really private societies or near-unions that are centered around food and drink. In many cases, steady employment in the hotel-food business is a prerequisite for membership, as well as some "knowledge" of the products involved—usually determined by qualifying examinations. However, individual members of these societies may be in a position to help the outsider. These groups all have local addresses for chapters. The easiest way to find their addresses is to contact a leading chef in your area, for he would have the necessary information. The following is only a partial list:

Les Amis du Vin
Caterina de Medici Society
Commanderie de Bordeaux
Confrérie de la Chaîne des Rôtisseurs
Confrérie des Chevaliers du Tastevin
Confrérie St. Etienne (Alsace)
Escoffier Society
International Wine and Food Society, London
Sommelier Society of America, 121 West 45th St., New York, N.Y. 10036

The next list is that of private individuals who usually earn their living from the alcoholic beverage trade, and who supplement their incomes by writing. Attorneys advise that no one here can properly establish an advisory service, since if they did so they would be responsible for the results. This is particularly in reference to amateur brewing. What they *can* do is provide descriptive guidance to any problems, make suggestions, or refer the reader to other printed literature. Most are open to correspondence through *Wine Magazine*, *Vintage*, and *Gourmet*. In any case, the living authors of the books described in this mediagraphy would certainly enter into worthwhile correspondence. Write your request in care of the publisher.

Asher, Gerald. Director of Imported Wines for Austin Nichols; author; wine columnist for *Gourmet*.

Bain, George. Former wine columnist for Toronto *Globe and Mail*; now with Toronto *Star*.

Balzer, Robert Lawrence. Author (Ward Ritchie Press, Los Angeles); wine consultant; wine columnist for the Los Angeles *Times*.

Barbour, Beverly. Syndicated columnist ("Bev's Bits and Bites"); food consultant; contributor to *Vintage*.

Bassin, Barry. National wine sales manager for Dryfus, Ashby; lecturer; author of a 240-hour study course on wines.

Bayard, Luke. Writer and lecturer on wine; contributor to *Wine Magazine*.

Bidwell, Edward. Wine consultant with Sichel.

Bond, Jules. Food columnist for Southhampton *Press* and Suffolk (Long Island) *Times*; author of cookbooks; lecturer; officer of various societies.

Broadbent, Michael. Wine consultant for Christie's in London; author; contributor to *Vintage* and *Wine Magazine*.

Burke, K. C. Editor of *Wine Magazine*.

Chiapperini, Felice. Departmental Chairman, Hotel and Restaurant Management Program, New York City Community College (which was recently awarded a medal for culinary excellence by the French government).

Daniele, Mario R. Consultant with C. Daniele and Company, wine and food importers from France and Italy.

De Groot, Roy Andries. Author; contributor to *Esquire* and *Vintage*.

Feinberg, Harry. Importer (Monsieur Henri wines).

Garbani, Ferdinando. Director of Crosse and Blackwell Vintage Cellars (England).

Gerard, Whitney. Lawyer to the wine trade (clients: Schoonmaker, de Luze, Bass-Charrington); wine consultant to S.S. Pierce.

Gizelt, Charles. Vice President of Wine Food International (New York).

Gourdin, Robert. National director and founder of "Les Amis du Vin"; national sales manager of Schieffelin and Co.

Haas, Robert. Importer; agent for Austin Nichols.

Jeffs, Julian. Author; contributor to *Wine Magazine*.

Johnson, Hugh. Author; contributor to *Wine Magazine*.

Leedom, William. Author; assistant to Frank Schoonmaker (q.v.).

Lichine, Alexis. Vineyard owner; author (especially of the great *Encyclopedia of Wines and Spirits*); importer.

Lindquist, Lindy. West Coast editor of *Vintage*.

McNally, Alexander C. International wine manager for Heublein, Inc.; responsible for Heublein's annual wine auction.

Mariani, John, Jr. President of Banfi Products Corporation, largest importer of Italian, Swiss, and Austrian wines in the United States.

Massee, William. Author; wine columnist for the New York *Daily News*.

Misch, Robert. Syndicated columnist ("Eat, Drink and Be Merry"); author of the Quick Guide series; lecturer.

Mueller, Charles. Vice-president, director of wine sales for Kobrand Corporation; lecturer; chairman of various wines and foods committees.

Prial, Frank. Columnist (weekly) for the New York *Times*.

Rickard, Dominique. Wine Programs Director for *Vintage*.

Salvi, John. Contributor to *Wine Magazine*.

Schoonmaker, Frank. Author; importer; consultant to Almaden Vineyards.

Seldon, Philip. Publisher and editor of *Vintage*; co-ordinator of "Ask Vintage" column.

Sichel, Peter. Importer; vineyard owner; contributor to *Vintage*; author.

Waugh, Harry. Author; contributor to *Wine Magazine* and *Vintage*; consultant.

Wildman, Frederick, Jr. Author; consultant; former wine columnist for *Gourmet*.

Wile, Julius. Senior Vice-President of Julius Wile and Sons, importers; annual contributor to the *Encyclopaedia Britannica*; lecturer at Cornell's School of Hotel Administration.

Winroth, J. Columnist on wines and foods for the *International Herald Tribune* (Paris); an official wine taster for the French government.

Woon, Basil. Contributor to *Wine Magazine*; author of many "Report from . . ." columns detailing wine development throughout France.

Yoxall, H. W. Author; contributor to *Wine Magazine*.

The preceding list is just a starter. There are very few females writing about wines, beers, and spirits, although the editor of *Wine Magazine* (K. C. Burke) is a woman. Generally, women write about foods and wines. For further information on any aspect of wines, beers, and spirits, drop a line to either *Vintage* or *Wine Magazine*.

Chapter 11

DIRECTORIES

PUBLISHERS

Excluded from this section are publishers of periodicals, associations and clubs, and publishers of out-of-print material.

A.M.S. Press, Inc.
56 E. 13th Street
New York, New York 10003

AW Press (Amateur Winegrower)
South Street
Andover, Hants.,
England

Abelard-Schuman, Ltd.
257 Park Avenue South
New York, New York 10010

Academic Press, Inc.
111 Fifth Avenue
New York, New York 10003

Ahrens Publ. Co., Inc.
50 Essex Street
Rochelle Park, New Jersey 07662

Aldine Publ. Co.
529 S. Wabash Avenue
Chicago, Illinois 60605

American Elsevier Publ. Co.
52 Vanderbilt Avenue
New York, New York 10017

Arc Books, Inc.
219 Park Avenue South
New York, New York 10003

Arco Publ. Co., Inc.
219 Park Avenue South
New York, New York 10003

Arlington Books, Ltd.
38 Bury Street, St. James
London SW1 6AU, England

Avi Publ. Co.
Box 831
Westport, Connecticut 06880

Award Books
235 E. 45th Street
New York, New York 10017

J. B. Baillière at Fils
19 rue Hautefeuille
Paris 6e,
France

A. A. Balkema (Pty), Ltd.
Posbus 3117
Capetown, South Africa

André Balland
33 rue Saint André des Arts
Paris 6e,
France

A. S. Barnes and Co.
Forsgate Drive
Cranbury, New Jersey 08512

Barnes and Noble
10 E. 53rd Street
New York, New York 10022

Barrie and Jenkins
2 Clement's Inn
Strand, London WC2A 2EP
England

G. Bell and Sons, Ltd.
York House
6 Portugal Street
London WC2A 2HL
England

Adam and Charles Black
4-6 Soho Square
London W1V 6AD, England

Blakiston Books
Shoppenhangers Road
Maidenhead, Berks.,
England

Benjamin Blom, Inc.
2521 Broadway
New York, New York 10025

Bobbs-Merrill Co.
4300 W. 62nd Street
Indianapolis, Indiana 46268

The Bodley Head
9 Bow Street
London WC2E 7AL
England

Booknoll Farm
Hopewell, New Jersey

British Book Centre, Inc.
996 Lexington Avenue
New York, New York 10021

Cambridge University Press
32 E. 57th Street
New York, New York 10022

Cassell and Co., Ltd.
35 Red Lion Square
London WC 1
England

Chapman and Hall, Ltd.
11 New Fetter Lane
London EC4P 4EE
England

Chemical Publ. Co., Inc.
200 Park Avenue South
New York, New York 10003

Citadel Press, Inc.
120 Enterprise Avenue
Secaucus, New Jersey 07094

Clarendon Press
Oxford, England

Columbia University Press
562 W. 113th Street
New York, New York 10025

Constable and Co., Ltd.
10 Orange Street
Leicester Square
London WC2H 7EG
England

Consumers' Association
14 Buckingham Street
London WC2N 6DS
England

Consumers Union of the United States, Inc.
256 Washington Street
Mount Vernon, New York 10550

Cowles Book Co.
114 W. Illinois Street
Chicago, Illinois 60610

Crown Publishers, Inc.
419 Park Avenue South
New York, New York 10016

David and Charles, Ltd.
P.O. Box 4
South Devon House, Railway Station
Newton Abbot, Devon,
England

Peter Davies, Ltd.
15-16 Queen Street
Mayfair, London W1
England

Andre Deutsch, Ltd.
105 Great Russell Street
London WC1B 3LJ
England

Dodd, Mead and Co.
79 Madison Avenue
New York, New York 10016

Doubleday and Co., Ltd.
277 Park Avenue
New York, New York 10017

Dover Publications
180 Varick Street
New York, New York 10014

Drake Publishers
381 Park Avenue South
New York, New York 10016

E. P. Dutton and Co., Inc.
201 Park Avenue South
New York, New York 10003

Editions d'Art les Heures Claires (S.A.)
19 rue Bonaparte
Paris 6e, France

Editions Générales, S.A.
rue Gustav-Mognier 3
CH 1200 Genève, Switzerland

Editions Stock
14 rue de l'Ancienne Comédie
Paris 6e, France

Educator Books, Inc.
10 N. Main
Drawer 32
San Angelo, Texas 76901

Essandess Special Editions
630 Fifth Avenue
New York, New York 10020

Faber and Faber
3 Queen Square
London WC1
England

Farrar, Straus and Giroux, Inc.
19 Union Square West
New York, New York 10003

Finch Press Reprints
337 E. Huron Street
Ann Arbor, Michigan 48108

Follett Publ. Co.
1010 W. Washington Blvd.
Chicago, Illinois 60607

B. Franklin
235 E. 44th Street
New York, New York 10017

Funk and Wagnalls, Inc.
55 E. 77th Street
New York, New York 10021

Golden Press
850 Third Avenue
New York, New York 10022

Victor Gollancz, Ltd.
14 Henrietta Street
Covent Gardens
London WC2E 8QJ, England

Greystone Corporation
225 Park Avenue South
New York, New York 10003

Hamlyn Publ. Group
Hamlyn House
42 The Centre
Feltham, Middlesex,
England

Harper and Row
10 E. 53rd Street
New York, New York 10022

George G. Harrap and Co., Ltd.
182-184 High Holborn
London WC1V 7AX
England

Hart Publ. Co.
719 Broadway
New York, New York 10003

Hastings House Publishers, Inc.
10 E. 40th Street
New York, New York 10016

Hawthorn Books, Inc.
260 Madison Avenue
New York, New York 10016

William Heineman, Ltd.
15-16 Queen Street
London W1X 8BE
England

Hill and Wang
19 Union Square
New York, New York 10003

Hippocrene Books, Inc.
171 Madison Avenue
New York, New York 10016

Hodder and Stoughton, Ltd.
St. Paul's House
Warwick Lane
London EC4P 4AH
England

Holland Press, Ltd.
112 Whitfield Street
London W1
England

Holloway House
8060 Melrose Avenue
Los Angeles, California 90046

Horizon Press
156 Fifth Avenue
New York, New York 10010

Horwitz Group Books Pty. Ltd.
Denison Street 2, P.O. Box 495
North Sydney, New South Wales
2060, Australia

Houghton Mifflin Co.
1 Beacon Street
Boston, Massachusetts 02108

Human and Rousseau
Box 4618
Capetown, South Africa

Indiana University Press
Tenth and Morton Streets
Bloomington, Indiana 47401

International Publications Service
114 E. 32nd Street
New York, New York 10016

Michael Joseph, Ltd.
52 Bedford Square
London WC1, England

A. M. Kelley
305 Allwood Road
Clifton, New Jersey 07012

Alfred A. Knopf
201 E. 50th Street
New York, New York 10022

Lallemand
25 rue de la Faisanderie
Paris 16e, France

Lane Magazine and Book Co.
Menlo Park, California 94025

Librairie Damidot
13 rue des Forges
Dijon 21, France

Librairie E. Flammarion et Cie
26 rue Racine
Paris 6e, France

Librairie Hachette
79 boulevard Saint Germain
Paris 6e, France

Librairie Larousse
17 rue du Montparnasse
Paris 6e, France

J. B. Lippincott Co.
E. Washington Square
Philadelphia, Pennsylvania 19105

Little, Brown and Co.
34 Beacon Street
Boston, Massachusetts 02106

McCall Books
201 Park Avenue South
New York, New York 10003

McGraw-Hill Book Co.
1221 Avenue of the Americas
New York, New York 10020

David McKay Co., Inc.
750 Third Avenue
New York, New York 10017

Macmillan, Inc.
866 Third Avenue
New York, New York 10022

Macmillan Publ. Co.
70 Bond Street
Toronto, Canada

The Mercier Press, Ltd.
4 Bridge Street
Cork, Eire

William Morrow and Co., Ltd.
105 Madison Avenue
New York, New York 10016

John Murray Publishers, Ltd.
50 Albemarle Street
London W1X 4BD
England

Nash Publ. Co.
9255 Sunset Blvd.
Los Angeles, California 90069

Thomas Nelson (Australia) Ltd.
597 Little Collins Street
Melbourne C1, Victoria
Australia

New American Library, Inc.
1301 Avenue of the Americas
New York, New York 10019

New English Library, Ltd.
Barnard's Inn, Holborn
London EC1N 2JR
England

New Jersey Agricultural Society
Box Y
Trenton, New Jersey 08607

New Press
553 Richmond Street West
Toronto, Canada

Newman Neame, Ltd.
6 Kirby Street
London EC1 8TU
England

Nitty Gritty Productions
Box 5457
Concord, California 94524

Northland Press
Box N
Flagstaff, Arizona 86001

Nourse Publ. Co., Inc.
8344 Melrose Avenue
Los Angeles, California 90009

Octopus Books
221 Park Avenue South
New York, New York 10003

Odhams Books, Ltd.
Hamlyn House
42 The Centre
Feltham, Middlesex,
England

101 Productions
834 Mission Street
San Francisco, California 94103

Oriel Press, Ltd.
32 Ridley Place
Newcastle-upon-Tyne, NE1 8LH
England

Pacific Coast Publishers
4085 Campbell Avenue at
Scott Drive
Menlo Park, California 94025

Pan Books, Ltd.
33 Tothill Street
London SW1
England

Penguin Books, Inc.
7110 Ambassador Road
Baltimore, Maryland 21207

Pergamon Press, Ltd.
Headington Hill Hall
Oxford OX3 OBW
England

Sir Isaac Pitman, Australia, Pty. Ltd.
Bouverie 158, P.O. Box 160
Carlton, Victoria,
Australia

Sir Isaac Pitman and Sons, Ltd.
Pitman House, Parker Street,
Kingsway, London WC2
England

Playboy Press
919 N. Michigan Avenue
Chicago, Illinois 60610

Pocket Books, Inc.
630 Fifth Avenue
New York, New York 10020

Popular Library
355 Lexington Avenue
New York, New York 10017

Prentice-Hall, Inc.
Englewood Cliffs, New Jersey 07632

Presses Universitaires de France
108 boulevard Saint Germain
Paris 6e, France

Putnam and Co., Ltd.
9 Bow Street
London WC2E 7AL
England

G. P. Putnam and Sons
200 Madison Avenue
New York, New York 10016

Pyramid Communications, Inc.
919 Third Avenue
New York, New York 10022

Quadrangle/The New York Times
Book Co.
10 E. 53rd Street
New York, New York 10022

Reynal and Co.
105 Madison Avenue
New York, New York 10016

Ward Ritchie Press
3044 Riverside Drive
Los Angeles, California 90039

Routledge and Kegan Paul, Ltd.
68-74 Carter Lane
London EC4V 5EL
England

Rutgers University Press
30 College Avenue
New Brunswick, New Jersey 08903

St. Anthony's Press
St. Anthony's Hall
Peasholme Green
York Y01 2PW
England

St. Martin's Press, Inc.
175 Fifth Avenue
New York, New York 10010

Scarecrow Press, Inc.
52 Liberty Street, Box 656
Metuchen, New Jersey 08840

Charles Scribner's and Sons
597 Fifth Avenue
New York, New York 10017

Sherbourne Press
1640 S. LaCienega Blvd.
Los Angeles, California 90035

Simon and Schuster, Inc.
630 Fifth Avenue
New York, New York 10020

Singing Tree Press
Gale Research Co.
Book Tower
Detroit, Michigan 48226

Société d'Edition les Belles Lettres
95 boulevard Raspail
Paris 6e, France

Souvenir Press, Ltd.
95 Mortimer Street
London W1N 8HP
England

Tom Stacey, Ltd.
28 Maiden Lane
London WC2E 7JP
England

Stackpole Books
Cameron and Kelker Streets
Harrisburg, Pennsylvania 17105

Stein and Day Publishers
7 E. 48th Street
New York, New York 10017

Tampalais Press
Box 1286
Berkeley, California 94701

Taplinger Publ. Co.
200 Park Avenue South
New York, New York 10003

Terry Publ. Co.
Box 525
Olympia, Washington 98501

Thorsons Publishers, Ltd.
37-38 Margaret Street
Cavendish Square
London W1N 8LT
England

Time-Life Books
Time and Life Building
Rockefeller Square
New York, New York 10020

Howard B. Timmins (Pty), Ltd.
Colophone House
68 Shortmarket Street
Capetown, South Africa

Transatlantic Arts, Inc.
North Village Green
Levittown, New York 11756

Tri-Ocean Books
c/o Western Book Service Co.
1545 Minnesota Street
San Francisco, California 94107

Charles E. Tuttle
28 S. Main Street
Rutland, Vermont 05701

Verlag Ullstein Gmbtt
Lindenstrasse 76
D1000 Berlin 61

Unipub, Inc.
Box 433
New York, New York 10016

University of California Press
2223 Fulton Street
Berkeley, California 94720

University of Oklahoma Press
1005 Asp Avenue
Norman, Oklahoma 73069

University of Texas Press
Box 7819
University Station
Austin, Texas 78712

University of Toronto Press
33 East Tupper Street
Buffalo, New York 14203

University of Washington Press
Seattle, Washington 98105

University Press of Kentucky
Lexington, Kentucky 40506

Van Nostrand Reinhold
450 W. 33rd Street
New York, New York 10001

Vanguard Press, Inc.
424 Madison Avenue
New York, New York 10017

Vista Books
Blue Star House, Highgate Hill
London N19
England

Walker and Co.
720 Fifth Avenue
New York, New York 10019

Wehman Brothers
158 Main Street
Hackensack, New Jersey 07601

Westover Publ. Co., Inc.
333 E. Grace Street
Richmond, Virginia 23219

Wine Advisory Board
717 Market Street
San Francisco, California 94103

Wine Art
Box 2701
Vancouver 3, Canada

Wine Publications
96 Parnassus Road
Berkeley, California 94708

Winepress Publ. Co.
4026 N. Longview Street
Box 4016
Portland, Oregon 97208

Wines and Spirit Publications, Ltd.
Victoria House, Vernon Place
London WC1, England

World Publ. Co.
110 E. 59th St.
New York, New York 10022

SOURCES OF OUT-OF-PRINT BOOKS

Some members of the Antiquarian Booksellers Association of America specialize in wine books, and they occasionally offer lists or catalogs.

Corner Bookshop
102 Fourth Avenue
New York, New York

Marian Gore
Box 433
San Gabriel, California 91775

The Old Book House
3217 West Cora
Spokane, Washington 99208

Elizabeth Woodburn
Booknoll Farm
Hopewell, New Jersey 08525

Ms. Woodburn offers Catalog 773, which describes over 1,000 items concerning wines, beers, and spirits (including related cookbooks), many of European origin. This 44-page booklet is updated from time to time. Ms. Woodburn also operates Booknoll Reprints, and she has republished John Adlum's *A Memoir on the Cultivation of the Vine in America* (1823), Agostin Haraszthy's *Grape Culture, Wine and Wine Making* (1862), and André L. Simon's *Bibliotheca Bacchia* (1935), among others.

SOURCES OF SUPPLY OF WINE- AND BEER-MAKING EQUIPMENT

Less than a decade ago there were few suppliers of wine concentrates, or of such wine- and beer-making supplies as dried hops, wine yeast, or pre-packaged and pre-measured chemicals (acid blend, pectic enzyme, or grape tannin). Home brewing and vinification drew upon the imagination of the individual brewer or vintner. Crocks, pans, natural fruits and yeasts, home grape presses, and the magical raisin were used, instead of today's readily available specialist supplies. Often, the source of these supplies is a corner drugstore, whose back corner shelf may display a selected number of concentrates, hydrometers, siphons, and yeasts. Or, it may be that complete wine supply section in a large department store. The mails, however, will also bring all the materials that any amateur or professional oenologist or brewer could desire. The following list is merely a sampling from the hundreds of specialty wine and beer suppliers who also offer mail purchase.

Aetna Wine Supplies, Inc.
708 Ranier Avenue South
Seattle, Washington 98144

The Bacchanalia
273 Riverside Drive
Westport, Connecticut 06880

Patrick N. Baker
17 St. John Place
Westbury, Connecticut 06880

Berg & Sons
511 Puyallup Avenue
Tacoma, Washington

Buze (Selemby)
Box 490
Norman, Oklahoma 73069

Bynum Winery
614 San Pablo Avenue
Albany, California 94706

The Chemical Rubber Company
2310 Superior Avenue
Cleveland, Ohio 44114

The Compleat Winemaker
P.O. Box 2470
Yountville, California 94599

Herters, Inc.
Waseca, Minnesota 56093

Hobby Winemaking, Inc.
2758 N.E. Broadway
Portland, Oregon 97232

Jim's Home Beverage Supplies
North 2613 Division
Spokane, Washington 99207

E. S. Kraus
P.O. Box 451
Nevada, Missouri 64772

La Pine Scientific Company
2229 McGee Avenue
Berkeley, California 94703

Milan Laboratories
57 Spring Street
New York, New York 10012

Nichols Garden Nursery
1190 North Pacific Highway
Albany, Oregon 97321

Party House Beverage
10420 16th Avenue S.W.
White Center, Washington

Presque Isle Wine Cellars
9440 Buffalo Road
North East, Pennsylvania 16428

Rockridge Laboratories
P.O. Box 2842, Rockridge Station
Oakland, California 94618

E. H. Sargent & Company
4647 West Foster Avenue
Chicago, Illinois 60630

Semplex of U.S.A.
P.O. Box 12276, Camden Station
Minneapolis, Minnesota 55412

F. H. Steinbart
526 S.E. Grand Portland Avenue
Portland, Oregon 97214

Student-Science Service
3313-15 Glendale Blvd.
San Francisco, California 94118

Van Waters & Rogers, Inc.
3735 Bayshore Blvd.
Brisbane, California 94005

Vino Corporation
Box 7498
Rochester, New York 14615

Wine Art of America
4324 Geary Blvd.
San Francisco, California 94118

Wine Barrel, Arrowhead Farms
Box 421, Patuxent Road
Gambrills, Maryland 21054

Wine Hobby
P.O. Box 428
Laguna Beach, California 92651

Winecraft Winery
8363 Center Drive
La Mesa, California 92041

The Winemakers Shop
Bully Hill Farms
R.D. 2
Hammondsport, New York 14840

The Wyne Table
P.O. Box 490
Norman, Oklahoma 73069

INDEX

Numbers in this author-title-subject index refer to entry numbers in the text, and *not* to page numbers. In an attempt to make this index more meaningful, important and/or neglected chapters of books have been brought out in the index itself (see, for example, the index entry under **Wine tastings**). Other emphases here are on geographic names, specific names of alcoholic beverages, and recipes.